What Your Colleagues Are Saying . . .

The Formative 5 was a must-read book for anyone teaching math. It provided a wealth of practical guidance for teachers to make good formative assessment part of their regular planning and instruction. In *The Formative 5 in Action*, Fennell, Kobett, and Wray take things to the next level, expanding the scope of the original book to include high school math and providing even more practical guidance, supplemented by video resources that give key insights from teachers and students and by illustrations of what great formative assessment looks like in real classrooms. Anyone teaching math in K–12 needs this book in their professional library. It's a game changer.

Dylan Wiliam
Emeritus Professor of Educational Assessment,
UCL Institute of Education
London, UK

The Formative 5 in Action captures the essence of humanizing assessment practices that celebrate students' strengths and extend their math thinking to build positive math identities for every learner. I love that these practical modules with videos, tools, and tech tips can be used in K–12 professional learning communities as well as in methods courses to engage teachers in analyzing student responses using the five techniques!

Jennifer M. Suh
Professor of Mathematics Education, George Mason University
Fairfax, VA

Fennell, Kobett, and Wray helped us grow our knowledge and use of formative assessment in their original book. They've taken it to a whole new level with this updated version by providing us with additional tools for implementation. The videos and classroom examples bring the book's ideas to life and aid us in better utilizing formative assessment in the mathematics classroom.

Kevin Dykema
President (2022–2024), National Council of Teachers of Mathematics
Eighth-Grade Mathematics Teacher, Mattawan Middle School
Mattawan, MI

Kudos to Fennell, Kobett, and Wray for expanding on their wonderful book, *The Formative 5*. I am impressed with their attention to the importance of feedback and the variety of techniques teachers can use to help students understand where they are going, be aware of what progress is being made, and address the next steps in the instructional progression. This book is a great tool for teachers who want to use assessment more effectively.

Kyndall Brown
Executive Director, California Mathematics Project
Los Angeles, CA

The Formative 5 in Action is the logical next step for building depth, making connections, and supporting a community of practice with teams of educators. The book unpacks the five assessment techniques in ways that are accessible for educators to incorporate into their teaching. Specifically, I appreciate how this book incorporates vignettes, technology, and videos to support the use of the techniques for classroom use.

Robert Q. Berry III
Dean and Professor, College of Education, University of Arizona
Tucson, AZ

The Formative 5 in Action illuminates the benefits, how-tos, and real-life stories around implementing formative assessment techniques in the K–12 classroom. The interactive and user-friendly modules create a crucial resource expanding upon the first edition and connecting theory to realistic teacher practice. Additionally, the conversations around feedback and the importance of teacher reflection are impressive. I am excited to incorporate this resource into district work.

Desiree Y. Harrison
Elementary Instructional Coach, Farmington Public Schools
Farmington, MI

Formative assessment is one of the most powerful tools in a teacher's tool kit. This book beautifully presents five essential formative assessment techniques and breaks them down in clear language. Doing so makes formative assessment accessible to every teacher by using guiding questions, specific examples, and tools for reflection and analysis. I would use this book for self-study or in a professional learning community with my colleagues. It is a must for every mathematics educator's bookshelf.

Paul Gray
President (2021–2023), NCSM: Leadership in Mathematics Education
Dallas, TX

Accurate, authentic, and useful assessment is the linchpin of any K–12 mathematics classroom. When some people are asking, "What's the point of all this assessment?" Fennell, Kobett, and Wray are saying, "Learning is the point."
What's to love about this book? Among many things, it shares how to:

- Use students' strengths as the instructional starting line for the lesson.
- Reposition lesson closure from exit tickets to exit tasks.
- Deploy questions not just as probes but instead as potential pivot points that guide the direction of the lesson.
- Delve into students' thoughts and decision making while sharing targeted feedback.
- Move away from using feedback as suggestions for merely doing more or doing better.
- Use "you and me time" to create a private moment to build relationships and meet individual needs.

The Formative 5 in Action is *the* "playbook" for well-planned and easy-to-embed formative assessment tools that inform and as a result improve instruction and enrich learning. This is, in the authors' words, "important stuff."

Karen Karp
Professor, Johns Hopkins University
Baltimore, MD

The new features of this edition, including the strengths-based teaching, reflection tools, and videos, provide more engagement with the assessment strategies presented in each chapter. Fennell, Kobett, and Wray have teamed up to share practical approaches to assessment for mathematics educators at all levels.

Sandi Cooper
Professor, Baylor University
Waco, TX

In this new edition, the authors shout out, "Let's start with student strengths!" and follow through with a powerful expansion of formative assessment tools and resources. This will be valuable for K–12 teachers, instructional coaches, professional development leaders, administrators, district leaders, and many others. Throughout it is clear that planning, instruction, and assessment are inextricably linked to support student learning with intentionality.

Trena L. Wilkerson
Professor, Mathematics Education, Department of Curriculum and Instruction, Baylor University
Immediate Past-President, National Council of Teachers of Mathematics
Waco, TX

The Formative 5 in Action is a resource that all math educators must engage with. The concrete activities, videos, and planning tools provide clear guidance on how to use the five formative techniques, gracefully paired with feedback focused on the student's mathematical understanding. The reflective practice opportunities embedded throughout the book allow teachers to revisit their own practices and identify specific areas of growth with *The Formative 5*.

Ruby Norland
K–12 Mathematics Coordinator, Doral Academy of Nevada
Las Vegas, NV

This book includes concrete examples to center assessment for learning as integral to planning and teaching. The authors leverage the importance of feedback and observations to move student learning to greater depths of complexity. The reflections about assessment beliefs and habits provide a tool for centering discussions. What a wonderful tool for learning for teachers, coaches, and administrators!

Denise M. Walston
Chief of Curriculum/Director of Mathematics,
Council of the Great City Schools
Washington, DC

Fennell, Kobett, and Wray took their seminal work and made it more practical for the classroom teacher. As a current classroom teacher myself, I find the examples of student work, discussion around how to understand a child's thinking, and tips for planning next steps immeasurably beneficial as my colleagues and I engage in this work!

Zak Champagne
Lead Teacher and Director of Mathematics, The Discovery School
Jacksonville, FL

GRADES K–12

THE FORMATIVE 5 IN ACTION

GRADES
K–12

THE FORMATIVE 5 IN ACTION

Updated and Expanded From
The Formative 5: Everyday Assessment Techniques for Every Math Classroom

Francis (Skip) Fennell
Beth McCord Kobett
Jonathan A. Wray

A JOINT PUBLICATION

For information:

Corwin
A SAGE Company
2455 Teller Road
Thousand Oaks, California 91320
(800) 233–9936
www.corwin.com

SAGE Publications Ltd.
1 Oliver's Yard
55 City Road
London, EC1Y 1SP
United Kingdom

SAGE Publications India Pvt. Ltd.
Unit No 323–333, Third Floor, F-Block
International Trade Tower Nehru Place
New Delhi – 110 019
India

SAGE Publications Asia-Pacific Pte. Ltd.
18 Cross Street #10–10/11/12
China Square Central
Singapore 048423

Vice President and
Editorial Director: Monica Eckman
Associate Director
and Publisher, STEM: Erin Null
Senior Editorial Assistant: Nyle De Leon
Production Editor: Tori Mirsadjadi
Copy Editor: Melinda Masson
Typesetter: Integra
Proofreader: Dennis Webb
Indexer: Integra
Cover Designer: Scott Van Atta
Marketing Manager: Margaret O'Connor

Printed in the United Kingdom.

Library of Congress Cataloging-in-Publication Data

Names: Fennell, Francis M., 1944-, author. | Kobett, Beth McCord, author. | Wray, Jonathan A, author.
Title: The formative 5 in action, grades K-12 : updated and expanded from the formative 5 : everyday assessment techniques for every math classroom / Francis (Skip) Fennell, Beth McCord Kobett, Jonathan A Wray.
Description: Thousand Oaks : Corwin Press, Inc, 2023. | Includes bibliographical references and index.
Identifiers: LCCN 2023009317 | ISBN 9781071910559 (spiral bound) | ISBN 9781071913536 (epub) | ISBN 9781071913543 (epub) | ISBN 9781071913550 (pdf)
Subjects: LCSH: Mathematics--Study and teaching--Evaluation.
Classification: LCC QA11 .F462 2023 | DDC 510.71/2--dc23/eng20230415
LC record available at https://lccn.loc.gov/2023009317

This book is printed on acid-free paper.

23 24 25 26 27 10 9 8 7 6 5 4 3 2 1

Contents

Visit the companion website at
https://qrs.ly/wsetnnz
for downloadable resources, tools, and videos.

Note From the Publisher: The authors have provided video and web content throughout the book that is available to you through QR codes. To read a QR code, you must have a smartphone or tablet with a camera. We recommend that you download a QR code reader app that is made specifically for your phone or tablet brand.

"As a career-long early childhood educator, I never thought that much about assessing as I taught, but this formative assessment experience has helped me realize that I am always assessing! As I observe my students and informally interview them, I am assessing their progress and 'planting seeds' for my own planning and teaching."

—KINDERGARTEN TEACHER

"It has taken me a while to understand how important it is for me to anticipate what my students may do as they are engaged in doing the mathematics that I am teaching, and for me to use formative assessment to both monitor their responses and update their progress."

—FOURTH-GRADE TEACHER

"I have enough trouble with thinking about and planning for my teaching. How does my principal expect me to involve assessment, too—beyond our unit and end-of-year required tests?"

—SEVENTH-GRADE TEACHER

"Being honest, I always viewed assessments as my quizzes, end-of-marking-period tests, and our state-required assessment. I never learned that classroom-based formative assessments can truly link up with my planning and teaching. Thankfully, our math coach has helped me to not only understand but recognize the power of connecting my planning and teaching to my everyday use of formative assessment."

—EIGHTH-GRADE TEACHER

"I guess I was just lucky. My teacher preparation program included both coursework and practicum activities, which provided me with experiences where I needed to develop and use formative assessments connected with lessons I was actually planning and teaching. But what I did need to learn about and have experience with was feedback, including providing feedback to my students, as well as providing opportunities for them to provide feedback to me, and also to engage them in providing student-to-student feedback within my lessons."

—HIGH SCHOOL MATHEMATICS TEACHER

Preface

Throughout our years of experience in education—which has included classroom teaching, planning for and providing professional learning opportunities for teachers, and serving as mathematics leaders and teacher educators—we not only recognized but attempted to address many challenges related to assessment and, in particular, formative assessment. These include but are not limited to a lack of or limited awareness of the direct connection between planning, implementation, and assessment of instruction, as well as recognizing the well-documented evidence related to the impact of classroom-based formative assessment. And yes, we regularly confronted comments like those on the previous page. For close to a decade, while engaged in implementing the Elementary Mathematics Specialists and Teacher Leaders Project (www.mathspecialists.org), we regularly engaged and supported teachers, mathematics specialists, and mathematics teacher leaders. During this period, we consistently heard teachers, mathematics coaches/specialists, and mathematics teacher leaders talk about the challenges of assessment. They were quite serious about wanting to truly understand how formative assessment should impact the teaching and learning process. This was our tipping point, and truly prompted our own thinking and analysis of in-the-moment classroom-based formative assessment, which led to the publication of *The Formative 5: Everyday Assessment Techniques for Every Math Classroom* (Fennell et al., 2017).

While pleased with the response to and related success of *The Formative 5*, we knew that our work had just begun. In our experiences, which have included presentations and work in classrooms related to *The Formative 5* and assessment in general, we learned that to truly anticipate how they would use any of *The Formative 5* assessment techniques (e.g., Hinge Questions), teachers had to be deeply engaged in their use. We therefore decided that there was more to say, more to do, and more to show.

Like the original book (Fennell et al., 2017), *The Formative 5 in Action* represents a distillation and validation of particular classroom-based formative assessment techniques that teachers can use on a regular basis. We think of the Formative 5 as a palette of five "colors" that represent techniques teachers can use, sometimes mixing these colors to find the best way to formatively assess as well as guide planning, teaching, learning, and feedback opportunities every day. The *Formative 5 in Action* modules include an introductory module that identifies issues, challenges, and opportunities related to the importance and use of assessment in general and formative assessment in particular. This module also introduces readers to the modules dedicated to each of the Formative 5 assessment techniques: Observations (Module 1), Interviews (Module 2), Show Me (Module 3), Hinge Questions (Module 4), and Exit Tasks (Module 5).

WHAT'S NEW IN THIS BOOK?

You'll find a number of elements that have been added to this resource, building off of what educators found inspiring and useful in the original 2017 book.

1. **A new focus on feedback and strengths-based teaching and learning:** Like *The Formative 5: Everyday Assessment Techniques for Every Math Classroom* (Fennell et al., 2017), *The Formative 5 in Action* has been guided by the pioneering research of Dylan Wiliam and his colleagues (e.g., Black & Wiliam, 1998, 2009; Wiliam, 2018) and emphasizes the importance of minute-by-minute and day-by-day or short-cycle formative assessment and particular techniques for classroom-based formative assessment. We also recognized that we needed to provide a much stronger connection between use of the Formative 5 techniques and feedback, as well as make stronger statements about both identifying and starting with student strengths when engaging in and analyzing student responses to the techniques. Our links between strengths-based formative assessment and feedback in *The Formative 5 in Action* have been influenced by the research of John Hattie and Helen Timperley (2007) and more recent contributions by our own Beth Kobett and Karen Karp (2020).

2. **New and additional reflection, action, and implementation tools:** Each of the modules within *The Formative 5 in Action* is intended to function as a sort of playbook, truly engaging educators as they become familiar with suggestions for using a particular Formative 5 technique in the classroom. In addition, we offer many tools that can be used to guide planning, teaching, and recording of student responses for each of the Formative 5 techniques. You will also find blank versions of the tools that may be downloaded and adapted for your own use at the book's companion website, **https://qrs.ly/wsetnnz**. Finally, the last activity within each module of *The Formative 5 in Action* is titled "Your Turn," which seems like an appropriate phrase to launch *your* reading and engagement in *The Formative 5 in Action*.

3. **Video:** The book's modules contain videos related to each of the Formative 5 techniques to both demonstrate and validate particular techniques and jump-start your thinking about their use. In Modules 2 and 3, you'll also find quick-response (QR) codes that link to audio and video samples we have captured of students engaged in Interviews and Show Me classroom moments so that you can see their work and hear their thinking. Collectively, these are meant to demonstrate what these techniques both look and sound like in the classroom.

4. **Expansion to Grades K–12:** Finally, while we previously focused discussion of the Formative 5 techniques in Grades K–8, we saw the need to expand this discussion to address the formative assessment of mathematics teaching and learning to Grades K–12 and have added more discussion, teacher comments, and examples from high school mathematics to this volume.

So, think of this book, *The Formative 5 in Action*, as a more complete and more engaging opportunity to learn about and use Observations, Interviews, Show Me, Hinge Questions, and Exit Tasks in *your* classroom. We know that we will all continue to learn about the critical connections between planning, teaching, and assessment. We hope to continue to work with teachers at every level to determine our own next steps or additional tipping points as we continue to emphasize the power and importance of classroom-based formative assessment.

Acknowledgments

Our cumulative educational experiences include classroom teaching, serving in a variety of administrative and policy-related positions, serving as teacher educators, writing and analyzing research, creating book-length and journal manuscripts on a variety of topics, and providing professional learning opportunities for educators locally, regionally, and nationally. But the element of our journey together that has provided us with both our greatest challenge and a very high level of satisfaction, that some would call impact, has been our continuing efforts related to classroom-based formative assessment. We have believed for a long time that formative assessments and the anticipation of their use are essential components of teacher planning and instruction, and that such assessments guide and monitor the teaching and learning of mathematics. We continue to learn more about the importance of truly connecting assessment to planning and teaching, which includes the important role of teacher-to-student, student-to-teacher, and student-to-student feedback, as well as recognizing that any assessment technique must be implemented to spot student strengths as the first step in monitoring progress. We truly appreciate all those who provided their thinking and support as we developed *The Formative 5* (Fennell et al., 2017). Your comments relative to *The Formative 5* guided our learning and thinking as we created *The Formative 5 in Action*, which is both an update and an expansion and, we think, will truly engage readers in developing the confidence to use classroom-based formative assessment to guide and improve teaching and learning.

From Francis (Skip) Fennell: To Nita, Brett, Heather, and Stacey for their continuing support. To my former students, colleagues, and all others at Western Maryland College/McDaniel College for their encouragement and support. Special thanks and appreciation to those connected with the Elementary Mathematics Specialists and Teacher Leaders Project (ems&tl) who provided us with the tipping point that helped us define the Formative 5 and so much more related to our classroom-based formative assessment efforts. Special thanks and sincere appreciation to Beth Kobett and Jon Wray for your friendship and for allowing me to continue to learn—from you!

From Beth McCord Kobett: To my husband, Tim, and my daughters, Hannah and Jenna, for their constant love and support. Thank you to my Stevenson University teacher candidates, past and present, for sharing your beautiful strengths with me. You inspire and motivate me each and every day! To all the teachers, schools, leaders, and school communities that I have had the pleasure to work with and for, thank you for the many opportunities to learn and grow as an educator and human. To Skip and Jon, thanks for letting me join the project those many years ago on what turned out to be a life-changing adventure. I am forever grateful for this experience and your friendship.

From Jonathan (Jon) A. Wray: To my grandson, Parker, I love the way you always have a smile for me! And Julie, Alanna, Jordan, Annika, Molly, and Lorenzo—you are true blessings in my life. We are grateful for the ongoing support from professionals in the Howard County Public School System, around

Maryland, and beyond. We hope these techniques inform and impact your daily efforts to ensure that every student achieves mathematical excellence in an equitable, inspiring, engaging, and supportive environment. Skip and Beth, I am beyond thankful for our wonderful friendship and engaging work together!

From the authors: Words can't really express how much we appreciate our publisher at Corwin, Erin Null. Erin not only provided us with the freedom to test and craft the Formative 5, but also suggested we—make that *urged us to*—continue our journey with this update and significant expansion. *The Formative 5 in Action* is our next step in providing teacher support for understanding and implementing formative assessment techniques in the classroom—every day. Authors rarely get an opportunity to continue and extend their thinking, so we truly thank Erin for *The Formative 5 in Action.* We also appreciate the assistance of the following Corwin supporters: Senior Project Editor, Tori Mirsadjadi, for her editorial guidance; Nyle De Leon for her editorial support as she not only organized our seemingly endless manuscript flow but prepared our book for publication; and we are most appreciative of Melinda Masson's editorial expertise and Margaret O'Connor's marketing assistance as they guided the production of *The Formative 5 in Action.* Thank you!

PUBLISHER'S ACKNOWLEDGMENTS

Corwin gratefully acknowledges the contributions of the following reviewers:
Natalie Crist
Senior Manager of Content Development
The Math Learning Center
Salem, OR

Russell Gersten
Executive Director of Educational Research Institute
Instructional Research Group
Los Alamitos, CA

Ruth Harbin Miles
Retired K–12 Supervisor of Mathematics
Olathe District Schools
Staunton, VA

Christopher R. Horne
Curriculum Specialist for Elementary Science
Frederick County Public Schools
Frederick, MD

Karen Karp
Professor, School of Education
John Hopkins University
Baltimore, MD

Angela Waltrup
Content Specialist, Elementary Mathematics
Washington County Public Schools
Hagerstown, MD

About the Authors

Francis (Skip) Fennell, PhD, is emeritus as the L. Stanley Bowlsbey Professor of Education and Graduate and Professional Studies at McDaniel College in Maryland, where he also directed the Elementary Mathematics Specialists and Teacher Leaders Project (ems&tl). He is a former classroom teacher, principal, and supervisor of instruction, and past president of the Association of Mathematics Teacher Educators (AMTE), the Research Council on Mathematics Learning (RCML), and the National Council of Teachers of Mathematics (NCTM). He is a recipient of the Mathematics Educator of the Year Award from the Maryland Council of Teachers of Mathematics (MCTM), the Glenn Gilbert National Leadership Award from NCSM: Leadership in Mathematics Education, the Excellence in Leadership and Service in Mathematics Teacher Education Award from the AMTE, the James W. Heddens Distinguished Service Award from the RCML, and Lifetime Achievement Awards from both the MCTM and the NCTM. In 2018, he received an honorary Doctor of Humane Letters degree from McDaniel College. Skip's many book-length and peer-reviewed journal publications and classroom experiences have focused on assessment, number sense, fractions, elementary mathematics specialists and mathematics teacher leaders, and teacher education.

Beth McCord Kobett, EdD, is a professor of education and associate dean at Stevenson University, where she leads, teaches, and supports early childhood, elementary, and middle preservice teachers in mathematics education. She is a former classroom teacher, elementary mathematics specialist, adjunct professor, and university supervisor. Beth also served as director of the First Year Seminar program at Stevenson University. She recently completed a three-year term as an elected board member for the National Council of Teachers of Mathematics (NCTM) and was the former president of the Association of Maryland Mathematics Teacher Educators (AMMTE). Beth leads professional learning efforts in mathematics education both regionally and nationally. Beth is a recipient of the Mathematics Educator of the Year Award from the Maryland Council of Teachers of Mathematics (MCTM) and the Johns Hopkins University Distinguished Alumni Award. Beth also received Stevenson University's Rose Dawson Award for Excellence in Teaching as both an adjunct and full-time faculty member. Beth believes in fostering a strengths-based community with her students and strives to make her learning space inviting, facilitate lessons that spark curiosity and innovation, and cultivate positive productive struggle.

Jonathan A. Wray is the coordinator of secondary mathematics for the Howard County Public School System. He leads the implementation of mathematics instruction and engages teachers and administrators in professional learning opportunities focused on equity-based and effective teaching practices and instructional leadership. Jon served as the project manager of the Elementary Mathematics Specialists and Teacher Leaders Project (ems&tl) and was an elected member of the National Council of Teachers of Mathematics (NCTM) board of directors. He is a former classroom teacher, teacher mentor, and past president of both the Association of Maryland Mathematics Teacher Educators (AMMTE) and the Maryland Council of Teachers of Mathematics (MCTM). Jon is the recipient of the 2020 Ross Taylor/Glenn Gilbert National Leadership Award for his significant contributions to mathematics education, concern for his fellow educators, self-reflection and continuous growth, and contributions toward addressing issues in mathematics education.

PART I

GETTING STARTED

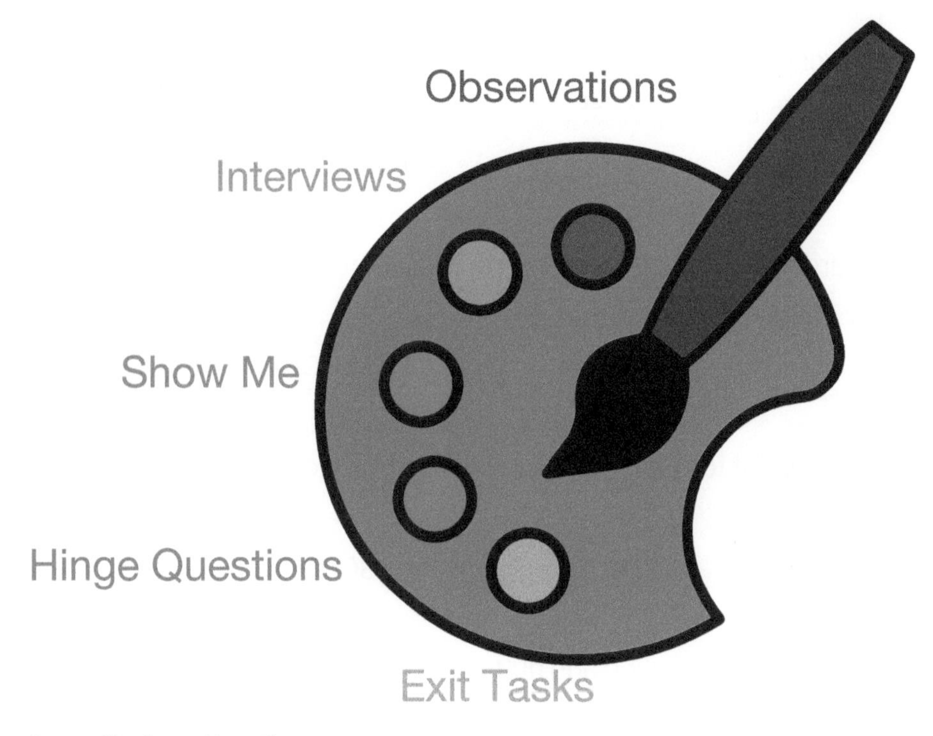

Source: iStock.com/Turac Novruzova

MODULE 0

WHY FORMATIVE ASSESSMENT? ISSUES AND OPPORTUNITIES

"I never really thought much about assessment other than the tests I would create and use or the end-of-year standardized required tests we used."

—KINDERGARTEN TEACHER

"When I first heard about assessment, I just figured they were talking about our end-of-year state-required tests."

—FIFTH-GRADE TEACHER

"Why didn't I learn about formative assessment in my teacher preparation program?"

—FIRST-YEAR MIDDLE SCHOOL TEACHER

"I just thought I could search online and buy whatever formative assessment I needed for math."

—HIGH SCHOOL TEACHER

FROM THE CLASSROOM

I admit to beginning my teaching career absolutely clueless with regard to any sort of "plan" about my use of and involvement with assessments. When it came time for me to do an end-of-topic test, I ran to my teacher colleagues and asked for help. Ever so helpful, they gave me tests they had created. I used them. Oh my. Did I know if such tests were formative or summative assessments? No! Did I consider what my students had done in our classroom and whether these "hand me down" tests were appropriate for what my students had learned and what I had taught? No! Without a doubt, I had a lot to learn.

I like to think—in fact, I know—I have come a long way. I know about the impact and importance of targeted summative assessments and classroom-based formative assessment. I can't imagine planning a lesson without thinking about observations, interviews, Show Me, hinge questions, and exit tasks to guide and monitor my teaching as well as provide me with incremental profiles of my students and their mathematics learning.

Purpose

An important everyday consideration of your planning and teaching is assessment. You control some of this! Classroom-based formative assessment monitors student progress, and also impacts your planning and instruction. That's on you, and it's what the Formative 5 is all about. However, summative assessment will also influence your teaching. Some of the summative assessments you may implement or analyze may be required by your school, school district, or state. Determining the influence of such external assessments on your teaching is important. As your planning and teaching intersect with both formative and summative assessment, the feedback you provide to your students, they provide to you, and students provide to each other identifies the instructional start line within a lesson or within the planning process.

Module Goals

As you read and complete activities within this module, you will:

- ✓ Understand the important role of assessment not only in measuring student progress, but also in monitoring and informing instruction.
- ✓ Differentiate between formative and summative assessment and reflect on the impact of each on your planning, instruction, and monitoring of student progress.
- ✓ Consider and reflect on your use of formative assessment as assessing to in*form*, with particular consideration of the five key strategies noted by Wiliam and Thompson (2008).

- ✓ Reflect on how regular, everyday use of classroom-based formative assessment informs your planning and instruction.
- ✓ Recognize the important connection between your assessments and the feedback you give to your students as well as the opportunities you will provide for students to provide feedback to you and to each other.

STUDENT LEARNING, TEACHING, ASSESSMENT, AND *YOU*: MAKING CONNECTIONS

Assessment of student learning is the responsibility of every school district, every school, and every teacher. Understanding and being able to use assessment to assist in teacher planning and instruction is an important element of a teacher's preparation and ongoing success in the classroom. Such assessment literacy includes being able to create, select, and effectively use classroom assessments and being able to select and effectively interpret and use results from external summative assessments.

Part of defining *your* assessment literacy means having the background and understanding to:

- Identify, select or create, and, of course, use assessments.
- Diagnose specific student instructional needs.
- Look for opportunities to focus on feedback: your feedback to students, their feedback to you, and student-to-student feedback.

This is essentially what this book is all about—classroom-based formative assessment. Important issues related to analyzing and evaluating the evidence generated by summative assessment, including externally developed and required summative assessments (e.g., annual state assessments), is really important, too, but that's another book for another time. Our focus will be on assessment as it relates, every day, to the classroom—*your* classroom.

Consider this module as the beginning of a journey that will start with an overview of particular issues and challenges related to assessment and then move to address, more directly, the classroom-based formative assessment techniques that are the focus of this book, and that you will use every day.

Video—Thinking About Assessment

Video 0.1
http://bit.ly/3ERts4N

To read a QR code, you must have a smartphone or tablet with a camera. We recommend that you download a QR code reader app that is made specifically for your phone or tablet brand.

Michele, Kristen, and Rebecca discuss their understanding and use of formative and summative assessment. Skip Fennell describes the intent of summative assessment and how formative assessment is connected to teacher planning and instruction, emphasizing the importance of ongoing connections between planning, instruction, and assessment.

Think about and discuss how closely your definitions and practice of formative and summative assessment align with or differ from how these ideas are discussed in the video. How closely is formative assessment connected to your teaching?

FORMATIVE, SUMMATIVE: IT'S ALL TESTING, RIGHT?

> "I actually never knew that my end-of-year and end-of-marking-period benchmark tests in mathematics were summative assessments. Thinking about how I can use both formative and summative assessments has been an eye-opening experience for me, *and* I'm in my fifth year of teaching!"
>
> **—FOURTH-GRADE TEACHER**

Assessment at the PreK–12 level has long been an assumed responsibility of the classroom teacher. You assess to determine and monitor student progress, to compare students, to guide and influence instruction, and to evaluate (e.g., evaluating a curricular program or instructional technique). Think about each of these purposes. When are you assessing to determine and monitor student progress? To compare students? To influence instruction? To evaluate? And, importantly, how much instructional time are you and your school district devoting to assessment? How are you using the assessment results—both assessments that you create and use and those that are external, typically summative assessments that you are responsible for administering (e.g., school district, state, or other mandated assessments)? Some argue, appropriately, that external summative tests are taking way too much time away from teaching and learning. For example, in *Testing More, Teaching Less* (Nelson, 2013), it was revealed that in one school district studied, students spent up to fifty-five hours per year taking tests. (That's about two full weeks of the school year.) One of the school districts studied had twelve different district and external summative assessments that accounted for forty-seven separate administrations of these assessments over the course of one instructional year.

For the classroom teacher, day-to-day involvement with assessment should be in the consideration and use of classroom-based formative assessments, while acknowledging the role and potential of summative assessment. Let's start this by considering, and defining, both formative and summative assessment.

Formative assessment has been discussed and seemingly defined and redefined for more than fifty years. Scriven (1967) and Bloom (1969) were early advocates of the power of formative evaluation to improve instruction. Based on their review of hundreds of studies, of which 250 were directly relevant to formative assessment, Black and Wiliam (1998) defined formative assessment "as encompassing all those activities undertaken by teachers and/or by their students, which provide information to be used as feedback to modify the teaching and learning activities in which they are engaged" (p. 7). The Council of Chief State School Officers (2018) defines formative assessment as "a planned, ongoing process used by all students and teachers during learning and teaching to elicit and use evidence of student learning to improve student understanding of intended disciplinary learning outcomes and support students to become self-directed learners" (p. 2). From our perspective, the focus of the Formative 5 is on the everyday use of classroom-based formative assessments to monitor, probe, and provide feedback designed to impact student learning and your planning and teaching.

INSIGHT
Formative assessment includes all activities that provide information to be used as feedback to modify teaching and learning.

What About You? Formative Assessment

Complete the following table and share your responses with a colleague.

How do you use formative assessment in mathematics?	**When** do you use formative assessment in mathematics?	**How** does your use of formative assessment influence your mathematics teaching?

Summative assessments are typically used to assess student learning at the conclusion of an experience. Such assessments could be a unit assessment, an assessment required by your school district, or the more high-stakes and high-profile end-of-year state assessments you may be required to administer. Many summative assessments are externally created—that is, prepared by others. Summative assessments are typically used to compare. Such comparisons could be student-to-student or class-to-class, or the extent to which results address predetermined standards or expectations.

Summative assessments are regularly used to identify score-based differences among individual students or among groups of students. These comparisons often lead to classifications of student scores on a student-by-student basis or on a group-by-group basis, using norms or defined levels of performance (e.g., *advanced, proficient, developing, not yet met*). It should be noted that summative assessment results or even performance on particular items of a summative assessment can also be used formatively when grade-level teams analyze results and use them to guide instructional goals and classroom activities. The Every Student Succeeds Act (2015) requires that all students complete a state-determined summative assessment in mathematics in Grades 3–8 and once at the high school level. However, states have flexibility in how and when they administer the tests (e.g., a single annual assessment can be broken down into a series of smaller tests). There's also an emphasis, beyond state-required assessments, on finding different kinds of summative tests that more accurately measure what students are learning. But most importantly, "if mathematics teachers are not drawn into discussions and decision-making about the use of summative assessments and their genuine influence on instruction, one wonders how we can continue to justify their overuse" (Fennell, 2020, p. 674).

What About You? Summative Assessment

Complete the following table and share your responses with a colleague.

What summative assessments are used at your grade level (elementary and middle school) or within courses you teach (middle/high school)?	**When** are they administered?	**How** are the results interpreted and used to influence *your* mathematics teaching and student learning?

INSIGHT

To summarize the differences between formative and summative assessment, many characterize summative assessments as **assessments of learning** *and formative assessments as* **assessments for learning**.

This book addresses a specific need regarding formative assessment, which is to identify and provide specific suggestions for how to use particular classroom-based formative assessment techniques on a regular—daily—basis. Our book is not about high-stakes summative assessments and the perceived, by many, overuse of such assessments. What we offer is designed to connect planning, teaching, and assessment in *your* classroom every day.

Time Out

The Importance of Assessment: What About You?

Let's Reflect

1. Does your school or school district have policies related to assessments and their use? If so, briefly describe the policies.

____

____

____

____

2. Are there opportunities for you to share formative assessments with members of your grade-level or departmental teaching team or with other school district colleagues? If so, what might you advise others regarding the sharing and related discussions of such assessments?

____

____

____

____

3. If there was one thing you could advise your school district's mathematics office about the use of assessments, formative and summative, what would that be?

____

____

____

____

FORMATIVE ASSESSMENT: ASSESSING TO IN*FORM*

> "I never really 'got' formative assessment. It just seemed to be like 'try this' and 'try that.' So many things to consider. Then I had this professor, and he used the painter's palette analogy. Small number of paints to choose from, which could be mixed and applied using various techniques and used daily. Got it. Worked for me! And now I use these classroom-based formative assessment techniques every day."
>
> **—MATH SPECIALIST/INSTRUCTIONAL LEADER**

Black and Wiliam (2009) noted that assessment becomes formative "to the extent that evidence about student achievement is elicited, interpreted, and used by teachers, learners, or their peers, to make decisions about the next steps in instruction" (p. 9). As stated previously, we know that formative assessment has been defined, redefined, researched, and discussed for decades. *Education Week* ("Understanding Formative Assessment," 2015) noted that formative assessment is both widely used and poorly understood! Some argue that the phrase *formative assessment* is open to too many interpretations. Stiggins (2005) and others actually prefer the phrase *assessment for learning*. Our position is that formative assessment is an integral component of what you do every day—planning and teaching—and that it involves a carefully defined and vetted set of assessment techniques specifically designed to in*form* instruction. Classroom-based formative assessment must engage students. How your students experience mathematics, which includes engagement in the assessment process, impacts the ways they identify themselves as knowers and doers of mathematics (National Council of Teachers of Mathematics [NCTM], 2018). As noted in *Principles to Actions: Ensuring Mathematical Success for All* (NCTM, 2014),

> *Effective teaching of mathematics uses evidence of student thinking to assess progress toward mathematical understanding and to adjust instruction continually in ways that support and extend learning. (p. 10)*

As a teacher, you are involved every single day in planning and teaching and then repeating that process. Assessment is integral to both planning and teaching. Linda Darling-Hammond (1994) noted that "in order for assessment to support student learning, it must include teachers in all stages of the process and be embedded in curriculum and teaching activities" (p. 25).

INSIGHT
Directly connecting assessment to planning and teaching within each lesson can truly influence teaching and learning.

Directly connecting assessment to planning and teaching within each lesson provides both the foundation and consistency in approach to truly influence teaching and learning. So, for instance, as you plan, consider not only mathematics content (e.g., place value) but also how the Standards for Mathematical Practice (National Governors Association Center for Best Practices & Council of Chief State School Officers, 2010) or mathematical processes (NCTM, 2000) will be integrally involved within a lesson. Linking assessment to planning in*forms* both teaching and learning (*form* within in*form* is italicized to bring attention to the central role of classroom-based formative assessment as it in*forms* teaching and learning).

Wiliam and Thompson (2008) suggest that the effective use of assessment for learning consists of five key strategies:

1. **Clarifying and sharing learning intentions and criteria for success with learners:**

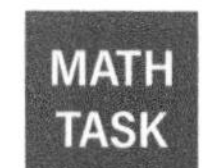

Jada cut 4 pieces of string, and each piece was 2 feet long. She placed the string pieces end-to-end. She thought she had more than 10 feet of string. Was she right? Can you show me how you decided if Jada was right or wrong?

The focus here is on unpacking the intended learning goals of a lesson and then determining the mathematical tasks and related activities that will lead to the expected learning. The example of the problem involving Jada provides a beginning task in multiplying whole numbers. The Show Me response requested (this formative assessment technique is the focus of Module 3) should demonstrate a level of understanding related to the mathematical intent of the lesson.

2. **Engineering effective classroom discussions, questions, and learning tasks that elicit evidence of students' learning:**

Using a drawing of rectangular or circular regions, show me three ways to represent fractions equivalent to $\frac{3}{4}$.

If we doubled the length of each side of a square, what would happen to the area of the square? What would happen if we tripled the length of each side of the square?

This assessment strategy considers how you will develop classroom activities that not only engage students in doing mathematics but provide evidence of student progress toward intended mathematical goals. The emphasis here is on the importance of taking the time to plan each lesson with a consideration for what you will assess and how you will assess the progress of your students. Think about what you would look for assessment-wise for each of the preceding examples. Careful planning—including attention to questioning, particularly the lesson's hinge question (more on that later)—and engineering the discussion of learning tasks address assessment *for* learning rather than *of* learning.

3. **Providing feedback that moves learners forward:**

"Great job! All five answers are correct."

"You solved the first three correctly. Look at problems 4 and 5 and see if you can find your mistake, and then show me how you would do these problems differently."

You provide feedback to your students each day. However, the most important thing about feedback is what students do with it. If your feedback prompted students to try a different solution strategy and they did so, then the feedback was helpful. Perhaps your feedback just affirms a student's response like the "Great job!" example. Whether or not specific feedback to your students "works" is really something that you can control. The more you observe your students as they engage in learning mathematics, the more you will get to know them and provide personalized feedback when they need it. In the example of doubling and tripling each side of a square, you may want to linger with the student so that the response can be quickly reviewed, and additional feedback provided as needed. See the next section of this module, which details how assessment and feedback are connected and linked to your planning and teaching.

4. Activating students as owners of their own learning:

"I like that pattern. Can you provide the next five numbers in the pattern and, as you do that, tell me why you have included them?"

"Show me how you know that the multiples and factors of a number are different."

Using formative assessment to monitor teaching and learning is not a one-way, teacher-to-student trip. The intent is to engage students in learning mathematics, which includes students taking an active role as they monitor and guide their own learning. One intent of formative assessment is to help students, all students, take an active role in and ownership of their learning. Such inclusive ownership and self-assessment opportunities will impact the pace of particular lessons and also have you consider particular formative assessment techniques. Your use of Observations, Interviews, Show Me, Hinge Questions, and Exit Tasks, the classroom-based formative assessment techniques presented in this book, will include the consideration of students as respondents, active learners, and fully engaged self- and peer assessors.

5. Activating students as instructional resources for one another:

Teacher: *Discuss and solve the next problem with your partner and be prepared to share the solution with the class.*

Cal: *When I looked at how Juan solved the problem, I really liked what he did. Next time I might try thinking about percent his way—rather than finding what you pay if it's 30 percent off, which is a two-step problem, thinking about the problem as 70 percent on (just subtracting 30 percent off mentally) turns it into just a one-step solution. I like that.*

Paired learning and small-group learning activities are instructional strategies you have most likely used throughout your teaching career. The formative assessment potential of peer review is in developing responsible collaboration among students. The result is that students learn from each other. Perhaps more importantly, students are often more willing to receive and accept feedback from a peer than an authority figure (e.g., teacher or parent/guardian) even when such student-to-student feedback is generally concisely presented and often very direct (e.g., "Why would you do it that way?" or "No way—that answer is not even close").

What we know about formative assessment is that student achievement can be improved when teachers regularly use it both within and between lessons. Our approach to classroom based formative assessment has been to focus on what Wiliam and Thompson (2008) have defined as short-cycle formative assessment—day by day and minute by minute. Our experience has been that such assessment is integral to and within every lesson, with the potential to impact students between lessons as well. While we recognize the importance of all the key strategies just discussed, our approach particularly emphasizes and promotes the following two strategies:

- Engineering effective classroom discussions, questions, and learning tasks that elicit evidence of students' learning
- Providing feedback that moves learners forward

INSIGHT
One intent of formative assessment is to help students, all students, take an active role in and ownership of their learning.

FOCUSING ON FEEDBACK

As you anticipate how your students will respond to a learning activity you are planning, you'll also need to think about the classroom-based formative assessments you will implement to monitor your students' progress. But wait. Planning, teaching, and using assessment to inform instruction without also providing, receiving, or facilitating feedback denies the importance of supporting student engagement. Feedback is not just connected to planning, teaching, and assessment; it's what defines the next steps learning-wise for your students and planning-wise for you. It is cyclical. Without this cycle, the learning–teaching continuity you so desire is lost.

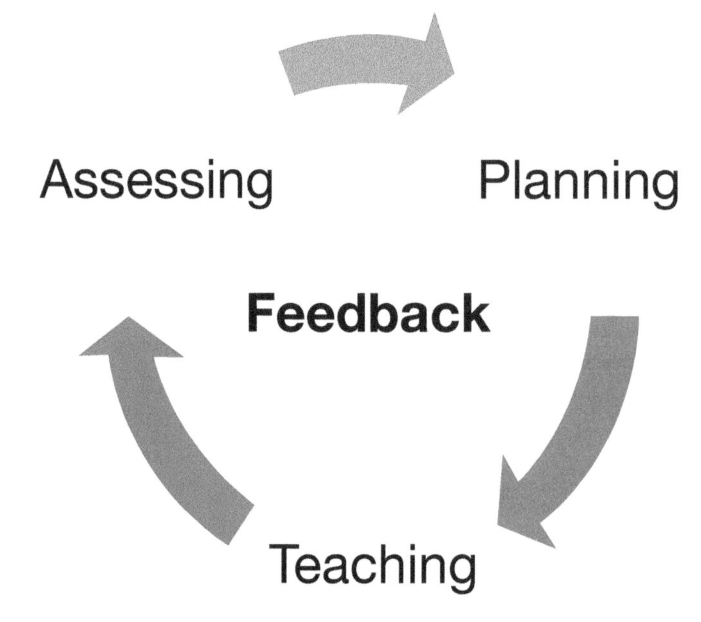

INSIGHT
Consider the relationships that you have built with your students and how these relationships inform the feedback that you provide. How do your students respond? What kinds of feedback best support your students?

Feedback is nuanced and must include attention to four attributes: (1) timeliness, (2) type, (3) purpose, and (4) *who* is receiving the feedback. The kinds of feedback teachers convey and seek—as well as how and when that feedback is delivered—can motivate students to work harder or, tragically, shut them down. When teachers know their students' mathematical, social, and emotional needs and find ways to cultivate positive relationships with those students, they can provide timely, explicit feedback that propels their students to work hard to seek mathematical understanding.

Our perception about the kinds of feedback we provide, and the timing of that feedback, may differ a bit from reality. Teachers can learn a lot about their feedback by conducting a feedback audit.

What About You? Feedback Audit

Ask a fellow teacher to observe you using the following chart and record only the feedback that occurs during the lesson. You could also simply audio-record a lesson or a portion of your lesson and then record the feedback.

Teacher-to-Student Feedback (Note the feedback and student the feedback is provided to)	**Student-to-Teacher Feedback** (Note the feedback and student who provided the feedback)	**Student-to-Student Feedback** (Note the feedback and students who provided the feedback)

As you review the feedback from your feedback audit, what do you notice?

- Does the feedback align with your perception? Why or why not?
- What is the overall quality of the feedback?
- Are there particular students who seem to receive more or less feedback?

Kobett and Karp (2020) propose the Feedback Loop model presented in Figure 0.1 (adapted from Hattie & Timperley, 2007), which illustrates the ordering and connections between determining and understanding expectations (*Feed Up*), receiving and responding to the responses and products of instruction (*Feed Back*), and next step actions you and your students will consider (*Feed Forward*). Kobett and Karp recognize the absolute necessity of planning and instruction that uses student strengths as the instructional starting line.

FIGURE 0.1 • **Strengths-Based Feedback Loop Model**

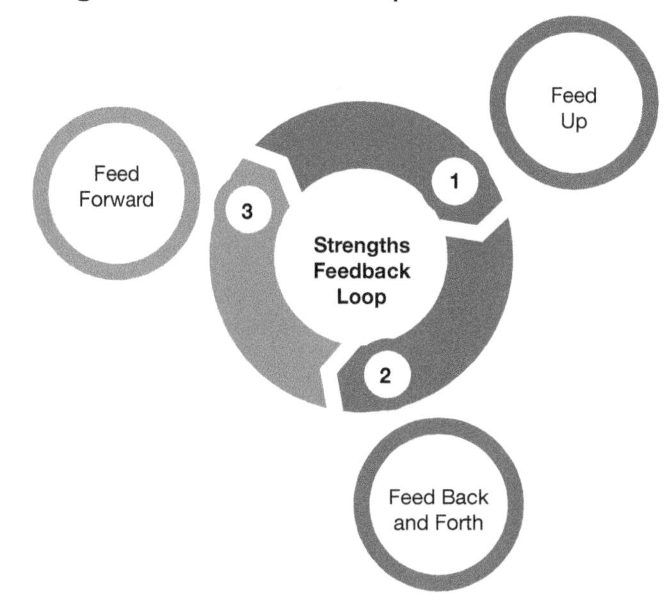

Source: Kobett & Karp (2020, p. 152).

Hattie and Timperley (2007) propose that feedback addresses three important questions you or one or more of your students may ask:

- **Where am I going?** (What's the intent, or the goals, of the learning activity?) *Feed Up* is about your students' understanding of the intent or goals of a learning activity. When your planning anticipates how you will assess students within the proposed learning activity, student understanding of what's expected learning-wise is more likely to occur.

WHAT "FEED UP" FEEDBACK LOOKS LIKE	WHAT "FEED UP" FEEDBACK SOUNDS LIKE
Teachers: • Refer to the goals for the lesson. • Remind students of what they have learned. • Show students how prior solution pathways have paved the way for new learning.	**Teachers say:** "Yesterday, we learned about ______. This new learning will help you today because ______." "You have been working in groups to solve problems together. Working collaboratively helps you to hear the mathematical reasoning of your peers. Today, I'd like you to ask your partners to share their ideas and compare their ideas to your own."

- **How am I doing?** (What progress is being made?) *Feed Back (and Forth)* is an indicator of your students' progress regarding a specific learning task. This may be your feedback to a student or class or feedback from a student, a group, or the class to you.

WHAT "FEED BACK AND FORTH" FEEDBACK LOOKS LIKE	WHAT "FEED BACK AND FORTH" FEEDBACK SOUNDS LIKE
Teachers: • Explicitly attend to students' mathematical understanding by affirming what students are doing, asking questions, and observing student responses. • Ask students to show what they are doing or perhaps how to extend a response or solve a related problem.	**Teachers say:** "The drawings you have created really show your thinking and the different solution pathways you tried!" "In your groups, share your solutions to the problem. Ask questions to learn more about how your partners solved the problem. Each of you should share your understanding of each other's solution strategy for the class problem." **Students say:** "I like that we could solve the problem in different ways. Using a drawing helps me get started."

- **What's next?** (What are the next steps, instructionally, as I/we focus on making progress?) *Feed Forward* addresses next steps, perhaps within the lesson—and, of course, in the planning, teaching, and assessment process—for the next day. This is the direct link to your use of classroom-based formative assessment—your Formative 5 feedback connection. Use of one or more of the soon-to-be-presented Formative 5 classroom-based formative assessment techniques and responses to the *Feed Up* and *Feed Back (and Forth)* questions defines *Feed Forward.*

WHAT "FEED FORWARD" FEEDBACK LOOKS LIKE	WHAT "FEED FORWARD" FEEDBACK SOUNDS LIKE
Teachers: • Extend student thinking by asking students to consider a new strategy. • Advance student understanding by prompting students to think in new ways about the mathematics content they are learning. • Use observation and interviews to design and adjust their lessons.	**Teachers say:** "What might be another method you can try to solve this problem? Which of those methods do you like best? Why?" "What would happen if we changed the number(s)? Would that make a difference? Why or why not?" "What might be a rule or formula that could work for other problems like this one?" **Students say:** "I tried two solution pathways and decided that _____ is the best because _______." "I am wondering if this method will work for other problems." "This problem reminds me of other problems we have solved because ______."

Let's consider your use of feedback. Sadly, when we discuss teaching and learning, assessment and feedback are often talked about later, often as an afterthought, as if assessment and feedback are not integral components of planning and teaching. A point that this book will make many times is the day-to-day connection between your planning, your teaching, the formative assessment techniques used to monitor student progress, and their direct connection to feedback—instructional goals understood by your students (*Feed Up*), how and when you will provide feedback to students during the day's mathematics lesson (*Feed Back and Forth*), and how your assessments and feedback will define your instructional next steps (*Feed Forward*), which could be in the moment within a lesson, as well as the starting point for your planning of the next day's mathematics lesson.

As noted, feedback is multidirectional and includes providing opportunities to encourage and receive student-to-teacher feedback relative to an assigned task or perhaps to a Show Me prompt (Module 3), as well as planning for opportunities for students, as they truly engage in the mathematics they are learning, to provide student-to-student feedback. We also recognize that the feedback provided or received actually launches the next lesson you will prepare, teach, and assess, confirming, once again, that assessing while you teach is what you do!

However, as you consider the feedback you provide, seek, or encourage your students to provide for each other, a most important consideration will be the actual focus of your feedback as it relates to student learning. Feedback is much more than a "great job" or "I like your graph" kind of comment, and feedback should not be implied as suggestions for just doing more or doing better. Also keep in mind that feedback is often thought of as teachers making statements about students, not about the impact and influence of their teaching (Timperley & Wiseman, 2002), thus diluting the impact and benefit of truly connecting assessment and feedback. The everyday use of and related student responses and feedback to formative assessment techniques are intended to guide and in*form* your everyday planning and instruction. Consider the activity that follows. Sort the feedback provided and respond to questions 1–3.

Focusing on Feedback: Activity

Sort the feedback statements for the following task.
Create a word problem for $\frac{1}{2} \div 4 = \frac{1}{8}$, representing the solution using a visual model.

A.	B.	C.
"Great job on your math drawing!"	"As I look at the problem you wrote—'Mabel has $\frac{1}{2}$ brownie and wants to share it with 4 friends. How much brownie will each friend get?'—and your model, which shows 4 brownies divided in $\frac{1}{2}$, I am wondering if they match."	"I see that many of you have written one word problem. Can you write another word problem and use a different model to solve it?"

D.	E.	F.
"Please review your solution—it doesn't make sense."	"I love how creative your math story is!"	"Your word problem—'Mary Alice is serving ice cream to her friends, but she only has 4 cups. How many $\frac{1}{2}$ - cup servings can she serve?'—is an interesting problem to showcase division of a whole number by a fraction. What model do you think will be most helpful to show your classmates?"
G.	**H.**	**I.**
"I noticed you used a region model to represent your word problem. Would a number line also work?"	"I am noticing that some of you used number lines and others used region models to show the division. Find someone who used a different model than you used and compare your word problems."	"I see that you have written a multiplication of fractions word problem. Take a look at this again."

1. **How did you sort the feedback?**
2. **What did you notice about the feedback?**
3. **Which of the teacher-to-student feedback comments provided opportunities for students to adapt, revise, and/or advance their thinking?**

While it seems natural to encourage students by making declarative statements about what is observed (Great job! Fabulous work! Nice model!), student thinking—and, often, working—stops. Providing explicit teacher-to-student feedback about what you observe during a task is a powerful way to reflect back to students how their learning demonstrates understanding.

CONSIDER THIS: MATHEMATICAL UNDERSTANDING AND FEEDBACK

Decisions about how we determine, interpret, record, and discuss student understanding are centered on the big idea that mathematical understandings are developmental. Important mathematical concepts and related procedures are developed over time and include many opportunities for students to grapple with the concepts or procedures being developed. Research has demonstrated that students must be able to make "connections between the new concept and at least two existing concepts" already understood (Yang et al., 2021). Recognizing that mathematical understanding is complex, we must carefully consider as we regularly engage in classroom-based formative assessment techniques both the ways we communicate what we assess and how we interpret students' mathematical understanding. Teachers frequently describe student understandings as misconceptions, implying that the student is wrong. But "from a child's perspective, it is a reasonable and viable response or conception based on their experiences in different contexts or in their daily life activities" (Fujii, 2020, p. 625). Student understanding that is resistant to change and persistent is often described as a misconception (Stacey, 2005). We recognize that all student mathematical understandings or conceptions range the full spectrum from small computational errors, to initial or partial conceptions, to fully developed conceptual understanding. But student understandings are just that—what students understand at any given time on any given day. Watson and Barton (2011) describe student conceptions as "developmental pathways" as students build greater mathematical understanding, adding to and revising previous understandings as they encounter new mathematical topics. Consider, for example, the mathematical conception that young students develop that "multiplying makes things bigger" and "dividing makes things smaller" as a mathematical idea that is developed through early experiences with multiplication and division of whole numbers and an example of early, naïve, or partial conceptions. Teacher use of language and use of "rules that expire"—something we tell students because it works right now, but later will no longer be true (Karp et al., 2014)—can heavily influence this kind of student conception.

A teacher's perception of student understanding greatly influences the feedback they will provide and the instructional decisions they will make. Student understanding that is early, or partial, requires a particular kind of feedback that provides the student with continued opportunities to grow in their understanding, while students with alternate conceptions might need to consider counterexamples to confront their mathematical ideas. Therefore, next we briefly describe student understanding and provide examples of related feedback.

Conception: Student understanding—whatever that might be at this point in time about the mathematics that they are learning. Feedback will likely collect more information from the student.
Examples of Feedback:

- Can you tell me more about your idea?
- Will that solution pathway work for another problem? Why or why not?
- I see that you are doing ________. That is a good idea because ________.
- What might happen if ________?
- Does your solution pathway make sense to you? Why or why not?
- Has your thinking changed about _______? Why or why not?

Mistakes: Typically, small errors.
Examples of Feedback:

- Does your answer make sense? Why or why not?
- I wonder if you could take a look at what you did again and see if you get the same solution.
- Could you and _____ compare your solutions?

Early, Naïve, or Partial Conception: Beginning or tentative understanding. Students may show inconsistent understanding. It is there one day, but not every day. This is normal and common in the learning process, as students grapple with their understanding.
Examples of Feedback:

- Tell me more about your mathematical idea.
- I wonder, if we tried this same method/strategy again, if we would get the same or a similar solution.
- Let's keep working on this idea you are developing with a new task.

Alternative Conception: Understanding that is built on prior ideas that worked before but don't work now. Or students applying an understanding that they have about another mathematical idea to a new mathematical idea.
Examples of Feedback:

- Tell me more about your solution.
- [Show students an example of the problem solved incorrectly (not their own work).] Is _____'s solution correct? Why or why not?
- [Pose two solved samples.] Which one makes the most sense? Why?
- I'd like you to pair up with ______ and compare your ideas.

CLASSROOM-BASED FORMATIVE ASSESSMENT: WHY IS THIS IMPORTANT? YOU DO HAVE THE TIME TO DO THIS!

> "It took me years to realize that assessment, particularly what I do involving classroom assessment, isn't some stand-alone 'other thing' I am supposed to be doing as required by my school's principal or supervisor. Hello, why didn't anyone tell me?!"
>
> —FOURTH-GRADE TEACHER

> "Formative assessment? I just thought it was something I was required to do."
>
> —KINDERGARTEN TEACHER

"I knew about formative assessment; it has just taken me a while to connect it to my classroom teaching and my planning."

—HIGH SCHOOL MATHEMATICS DEPARTMENT CHAIR

When you create, select, administer, and then evaluate the results of *any* assessment, formative or summative, you estimate the value of the responses and use that to determine what your students know. Important stuff. Probably by November of any given instructional year, maybe earlier, you have a sense of what each student in your mathematics classroom knows and is able to do. But the reality is, much of what you do assessment-wise is, or should be, directly related to what you teach—every single day. That's how we envision formative assessment. As noted earlier, the focus of this book is classroom-based formative assessment—the use of particular assessment techniques that you can and should use each day to not only validate and build on prior assessments, but also guide your planning and teaching. Why?

Consider NCTM's (2014) *Principles to Actions*:

> An excellent mathematics program ensures that assessment is an integral part of instruction, provides evidence of proficiency with important mathematics content and practices, includes a variety of strategies and data sources, and informs feedback to students, instructional decisions and program improvement. (p. 89)

The point here is that assessment must be an everyday component of what you do as you plan and teach. Assessing while you teach—it's what you do. You plan and teach, and as you teach a lesson, any lesson, you can—and should—use particular assessment techniques to monitor student progress within the lesson, as well as assessing the impact of the lesson itself.

In our early work with formative assessment, we recognized—and mathematics specialists and teachers told us—that there were so many suggestions and ideas related to formative assessment that understanding and using them was never well understood. And, in some cases, all the hype regarding formative assessment put the specialists and teachers on overload. This got our attention. We have spent time distilling and validating, through classroom use, a small set of classroom-based formative assessment techniques that teachers have used successfully on a regular basis. We like to think of these classroom-based formative assessment techniques metaphorically as a palette of five "colors" that you can use as you paint your own classroom canvas, sometimes mixing the colors to find the best way to formatively assess and guide teaching and learning on a daily basis. Later in this module, and much more specifically and in depth in the modules that follow, we will discuss the five techniques, which we call the Formative 5—Observations, Interviews, Show Me, Hinge Questions, and Exit Tasks.

Let's think about how formative assessment links to your own daily and even long-term instructional planning. As you plan, you consider the mathematical focus of a lesson. An important prerequisite to such planning is your own understanding of the mathematical content and pedagogical knowledge related to your grade level and beyond. We fully recognize that it takes time for you to understand the learning trajectories of the mathematics content topics for which you are responsible, as well as how to interpret and address them in your classroom. For example, for a lesson at the fourth-grade level related to equivalent fractions, some of your students may be able to move quickly into lesson extensions involving comparing and ordering fractions, while others may have difficulty representing common equivalent fractions. As you know, such an achievement range within a single mathematics topic is not uncommon. However, your ability to plan based on knowledge of your students and their mathematical needs is important. This certainly includes particular tasks you may select and design to match your lesson's mathematical focus and, importantly, how you will assess student performance and the overall impact and effectiveness of your lesson. In short, as you plan, you should anticipate what you expect your students to accomplish. So, yes, what and how you will assess is part of both planning and teaching. Your teaching will reflect the formative assessment techniques you have planned to use to monitor student progress and the lesson's overall effectiveness. The following questions may help guide your planning and teaching as connected to your use of formative assessment:

INSIGHT
Assessing while you teach—it's what you do.

- What tasks and questions will be used to engage students in the lesson?
- How will learning trajectories of the mathematics content focus of the lesson be considered to ensure the developmental appropriateness and student prerequisite background for this lesson?
- How will you communicate student learning expectations for this lesson?
- When and how will students receive feedback for their contributions during the lesson? And, how and when will students provide feedback to you?
- What responsibilities do your students have for assessing each other's learning in this lesson?
- How will formative assessment be used to monitor student progress in this lesson?
- Will students be assessed individually, in groups, or both individually and within a group?
- How will formative assessment be used to determine the effectiveness of the lesson?

Now it's time to consider the specific classroom-based formative assessment techniques that you can use in your classroom.

FORMATIVE ASSESSMENT IN *YOUR* CLASSROOM: THE CLASSROOM IS *YOUR* CANVAS!

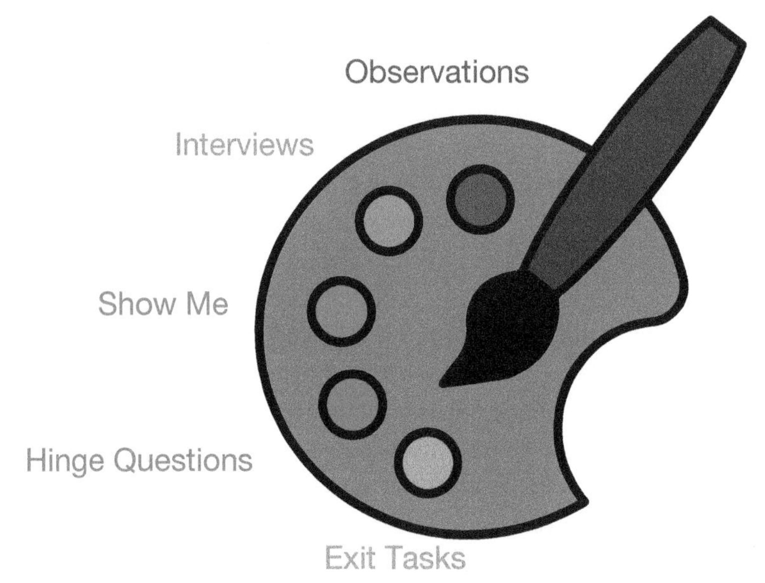

Source: iStock.com/Turac Novruzova

> "I never thought much about using formative assessment every day and had no idea how it connected with my planning and teaching. So glad we decided to use observation, interviews, Show Me, hinge questions, and exit tasks regularly. I get it now, and my kids have actually come to expect the hinge question and exit tasks."
>
> **—THIRD-GRADE TEACHER**

As noted earlier, this book presents five classroom-based formative assessment techniques, the Formative 5, which you can use every day. Using the metaphor of an artist's palette of five colors, the assessment techniques can be ordered and mixed based on your planning and instructional needs. The modules that follow will present, discuss, and provide tools for using the Formative 5 techniques—Observations, Interviews, Show Me, Hinge Questions, and Exit Tasks—in your classroom. This palette of formative assessment techniques has been gleaned from the seemingly endless suggestions provided for classroom consideration and use and has been carefully defined and tested in classrooms. A brief summary of each of the Formative 5 techniques is provided as follows.

INSIGHT
The palette of formative assessment techniques can be ordered and mixed based on your planning and instructional needs.

Observations

You observe your students every day—throughout the day. While this technique may be the most informal classroom-based formative assessment, its use is of particular importance to you as you monitor a lesson. As you use observation as a classroom-based formative assessment technique, the following questions, which will be discussed in depth in Module 1, will be helpful as you plan for the use of this technique.

1. What will you expect to observe?
2. How will you know "it" if you see it?
3. What particular strengths or challenges might you observe?
4. How will you record and provide feedback of what you observe?

A major intent of the Observations module (Module 1) is to provide the background and support tools that should assist you in using observation as a formative assessment technique to guide and in*form* your planning and teaching and monitor student progress.

Interviews

An interview extends an observation. The Interviews and Observations techniques are almost always connected. An interview provides the obvious follow-up to an observation a teacher might make when implementing a lesson. An interview also allows the teacher to spend a few valuable minutes digging deeper with an individual student or perhaps a small group of students. The goal of the interview is to get a glimpse of what a student is thinking. A full discussion of the Interviews technique, including helpful interview tools, is provided in Module 2. The following questions, also presented in Module 2, should help guide your use of this technique.

1. What would make you decide to work with a student one-on-one or with a small group of students?
2. What interview questions might you ask? How might the questions be different for particular students?
3. What responses will you anticipate from students? (Consider understandings *and* possible challenges.)
4. What follow-up interview questions might you ask, and how would such questions be connected to the feedback you might provide to the student or group of students?

The Interviews module will provide you with the background and tools appropriate to conduct, analyze, and use interviews to both monitor student progress and guide your planning and teaching.

Show Me

Show Me is a performance-based response by a student and, like an interview, extends an observation. Show Me occurs when a student, a pair of students, a small group, or perhaps the entire class is asked to show how something works, how a problem was solved, how a particular manipulative material or related representation was used, and so on. Teachers and mathematics leaders who have used the Show Me technique have noted that it validated information gathered from an observation and/or interview and often provided the first step in redirecting student responses. The following questions have proven to be helpful when anticipating use of the Show Me technique.

1. How is your Show Me different from an observation and interview?
2. What will you use as a prompt for a Show Me request for this lesson?
3. What might you want a student or students to show and say as they describe their Show Me response?
4. Recognizing that a student response to a Show Me prompt is student-to-teacher feedback, when would you provide teacher-to-student feedback to a Show Me response?

Module 3, the Show Me module, provides a full discussion and includes related tools useful for presenting and using the Show Me technique. The Observations, Interviews, and Show Me techniques are all quite connected. You will use each of them every day, with the Observations typically, but not always, helping to define the specifics of the Interviews and Show Me opportunities.

Hinge Questions

The hinge question (Wiliam, 2011) provides a check for understanding/proficiency at a "hinge point" in a lesson. The hinge question is a question that you plan for and use to elicit responses indicating your next step planning-wise and instructionally, with particular implications for the next day's lesson. Responses to the hinge question directly in*form* both planning and instruction.

Creating the hinge question is an important part of the planning of the day's lesson. Our experience has been that teachers need to take the time to create a question that truly assesses a major focus of the day's lesson. We often consider the hinge question as the lesson's "deal-breaker" since responses help you to determine your next steps instructionally, perhaps within the lesson you are teaching. We have also found it helpful to actually try out hinge questions with colleagues within a grade-level or departmental professional learning community. Such trial opportunities also provide teachers with occasions to consider varied hinge question formats. Most importantly, your ability to engineer the use of the hinge question is critical. Considering how you will engage students, assess responses, provide feedback, and decide instructional next steps attests to both the value and importance of the hinge question. Suggestions for the use and types of hinge questions are presented in Module 4. This will be a particularly important module for you.

Exit Tasks

The NCTM's (2014) *Principles to Actions* emphasizes the importance of using tasks to elicit student learning and then using the resulting analysis to inform instruction. We consider exit tasks as end-of-lesson formative assessments. We deliberately define such assessments as exit tasks given our experience with the seemingly increasingly popular use of exit tickets or exit slips. The exit task is designed to provide a capstone problem/task that captures the major focus of the mathematics lesson for that day or perhaps the past several days. The use of such problem-based tasks is quite different from the exit tickets or exit slips we have reviewed that tend to address particular mathematical procedures or provide opportunities for students to rate their level of understanding on the mathematics topic of the day. The exit task is actually a product, providing actual work samples for you to review and use for future planning. As for the hinge question, planning time will be needed to develop the exit task, and such task development is enhanced when school, department, or grade-level teacher learning communities work together in their creation, use, and revision. Questions to consider in exit task development include, but are not limited to, the following:

1. Does the exit task capture the mathematics content expectations of your lesson?
2. Given the grade level or mathematics course, classroom norms, and students' prior experience working with challenging mathematical tasks, will this exit task engage all of your students?
3. Should the exit task be completed by individual students, student pairs, or small groups?
4. When will you be able to review exit task responses and use the responses to guide your planning as well as provide feedback to your students?

The Exit Task module (Module 5) includes multiple examples of exit tasks and tools guiding their use. Given the performance and product nature of the exit task, it is not likely that you will use the exit task each day. Our experience has been that teachers use exit tasks two or three days per week and that the student responses guide not only daily but longer-term planning.

CONSIDER THIS: FORMATIVE ASSESSMENT AND GRADING

The focus of this book is the understanding and use of classroom-based formative assessment, with the intent to inform your practice by guiding your planning and teaching every day and by providing a constantly updated profile of individual student and class progress in learning mathematics. Should you grade responses to any of the Formative 5 techniques? While our quick response is no, that's not the intent of the response. Grades symbolically represent what you are constantly updating every minute of every classroom day—student progress. Consider the following: Would you individually grade what you observe, the interview responses of a student, or feedback to a Show Me response, hinge question, or exit task? We think not. Our position about grading student responses to any of the Formative 5 techniques you will learn about and use has been framed by our concern that grading when students are initially learning a particular concept or topic is not informed grading. At this point, learning is developing and emerging as students are observed or interviewed, respond to a Show Me request, answer a hinge question, and even work through an exit task. Evaluating students as they are just learning about, exploring, and engaging with new concepts does not give a complete picture of their understanding and may contribute to student anxiety or even fear or a reluctance to perform. Additionally, grading a formative response to any of the techniques presented in the next five modules sends a message to students that they are expected to immediately "get" particular concepts/skills/understandings.

We do know that some teachers with whom we have worked have found a workable way to grade exit tasks. Some have even graded Show Me responses. Interestingly, both techniques typically provide a performance-based written response. The decision to grade student responses to an exit task or any other formative assessment is yours, but the determination, framing, and use of the Formative 5 is more about embedding assessment with your planning and teaching to seamlessly address student progress and, for the reasons already noted, was never considered a formal grading opportunity.

SUMMING UP

Consider the title of this module: *Why Formative Assessment? Issues and Opportunities.* Even without thinking about it, you assess student progress all day long, every day. You observe, you talk to your students about what they are learning, you ask students to show you what they are doing, you ask questions, you provide feedback, and so much more. The module started by discussing the importance of assessment, and we recognize that such understanding is the foundation to truly recognizing the importance, value, and use of both formative and summative assessment and how, in particular, formative assessment can and must guide and monitor your teaching every single day. That's what this module and the following modules are all about—understanding the opportunities related to classroom-based formative assessment and considering how you can make a difference as you connect your planning and teaching to particular classroom-based formative assessment techniques—the Formative 5. This palette of formative assessment techniques—Observations, Interviews, Show Me, Hinge Questions, and Exit Tasks—was presented briefly in this introductory module, and will be thoroughly discussed and analyzed, with lots of tools supporting its use, in subsequent modules. The Formative 5 techniques represent our response to addressing issues and opportunities related to formative assessment. Let's get started.

Your Turn

Rate, Read, Reflect! Consider the following questions. Complete and then discuss your responses with your grade-level (elementary, middle school) or department (high school) teaching team or with teams across multiple grade levels.

1. Your school or school district's policies related to formative and summative assessment are known and understood by most teachers.

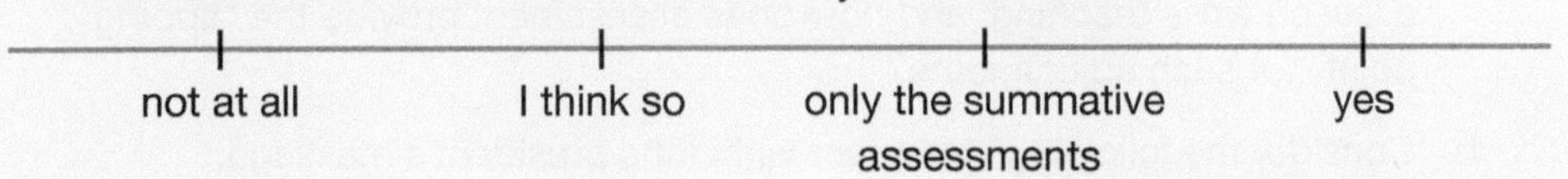

2. How often do you actually plan for your use of formative assessment and the related feedback to your students?

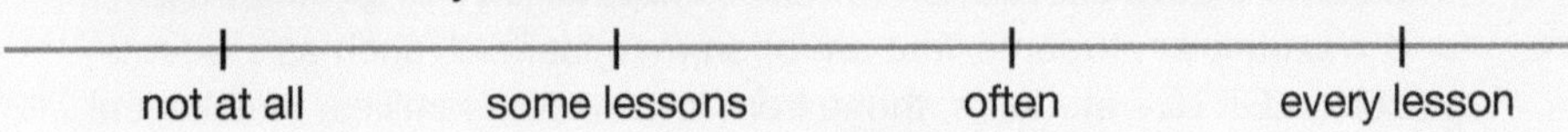

3. When you plan a lesson, do you anticipate what you will assess and when and how you will provide feedback to your students?

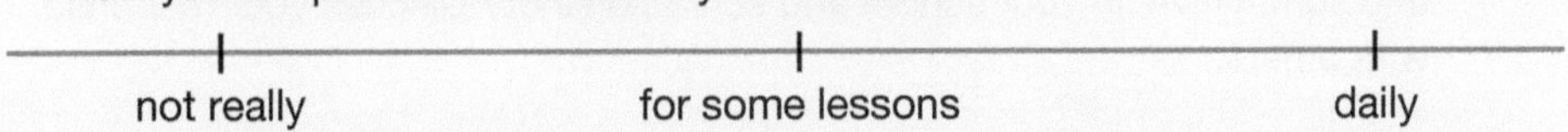

4. How does (or perhaps should) the use of formative assessment influence your instructional planning?
5. How much time do you spend each day assessing student progress in mathematics?
6. How much time do you spend each month and during the entire school year assessing your students? Make sure to include the summative assessments you administer as well as your use of formative assessments.
7. In your own words, describe the differences between formative and summative assessments.
8. What formative assessment techniques are you currently using?
9. How do you provide feedback to your students with regard to the assessments that you use? What opportunities are provided for students to provide feedback to you and for students to provide feedback to each other regarding their mathematics learning?

(Continued)

(Continued)

10. What concerns you the most about the imbalance, particularly as emphasized in reporting to parents/guardians and in the media, between formative and summative assessments?

11. Professional Connection: Consider reading *Using Formative Assessment Effectively* by NCTM president Trena Wilkerson (2022), available from https://www.nctm.org/News-and-Calendar/Messages-from-the-President/Archive/Trena-Wilkerson/Using-Formative-Assessment-Effectively/, then discuss the following:

    a. Wilkerson discusses "short cycle" assessments (Wiliam, 2018) and how they help teachers respond in real time to the learning needs of students. What is it that helps teachers get to the point where they can literally adjust a lesson while teaching, and how does assessment provide the "tipping point" for such adjustments?

    b. Consider the following statement within the president's message: "Too often students are marginalized through ineffective, inequitable assessment practices, including formative assessments. We must ensure that our formative assessment practices are inclusive of all students, and in particular students who are often marginalized, such as students of color, LGBTQ+ students, those from low-income families, multilingual students, and students with disabilities" (Wilkerson, 2022). Would this statement represent the inclusive nature of assessments, both formative and summative, in your school and school district? Discuss your response with others.

PART II

THE FORMATIVE 5 IN ACTION

MODULE 1

OBSERVATIONS

"How is observation an assessment? Of course, I observe my students—all day long, every day! I just never considered the assessment potential of my observations!"

—FIRST-GRADE TEACHER

"I actually know more about my students because I am always watching them work and seeing how they interact—with the mathematics they are learning and with each other. For me, observation is my everyday formative assessment lifeline!"

—FOURTH-GRADE TEACHER

"I try to regularly (once a week or so) write a brief summary of my observations of each of my classes. These 'weeklies' are really helpful, and as I look at the summaries, I am regularly amazed at the progress of individual students and of my classes."

—HIGH SCHOOL MATHEMATICS TEACHER

FROM THE CLASSROOM

I've taught for fifteen years and served the last three years as my school's math specialist. If you had told me, early in my career, how important observations were as a formative assessment technique, I am not at all sure I would have understood what you were talking about.

Now, as I work with my teachers, we talk, regularly, about the power of observations and how important observation is as a formative assessment technique. Several of my teachers have commented that they have always observed their students all day long, every day, but now they observe differently.

Marny was saying just the other day that when she plans her lessons, she thinks about—actually anticipates—what she might observe her students doing and what "it" might look like. She is hoping for certain responses, but also thinking realistically about how particular students may respond, and planning for such expectations, now. She thought, and others in her grade-level group agreed, that with our emphasis on formative assessment, her planning is more anticipatory.

Yes, observations are really important. I learned years ago that what I observe helps me monitor the progress of my students and of my instruction; it's gratifying that my teachers see the value of observations and recognize that what they have been doing their entire career, observing students, is so critical to their planning, teaching, and assessing.

Purpose

No matter where you are, as you are teaching, you are always completing an observation. You may just be scanning the room to see what everyone is doing, or you may be watching for something specific. In the classroom, as part of the observation process, you watch, notice, and consider how you will use your findings. You will now start to think about your own planning and the use of an observation as a formative assessment technique.

Module Goals

As you read and complete activities within this module, you will:

- ✓ Recognize and understand how you will use observation as a formative assessment technique to guide your planning and monitor your teaching.
- ✓ Plan for observations focusing on the importance of anticipating as part of the planning process.
- ✓ Conduct observations in your classroom.
- ✓ Reflect on how you have used observations and how use of observations has helped you determine the instructional needs of your students.
- ✓ Create a plan for how you will incorporate the use of observations, every day, as you plan and teach mathematics.

OBSERVATIONS: BACKGROUND AND BASICS

As you consider classroom-based formative assessments, the Observations technique is your "first stop" in the everyday application of the Formative 5. By **observation**, we mean directly observing student and class progress on particular mathematics activities. You watch, and you notice, but the intent here is to consider how you will use daily observations to inform your planning and teaching. In our work, we often find that Observations, while perhaps the most informal and readily used of the five formative assessment techniques presented (Modules 1–5), is both taken for granted and, at least to an extent, the least understood of the five techniques.

INSIGHT
Observing is what you already do every day, all day long.

Observations have been supported as a research-based practice for decades. Freudenthal (1973) indicated that "we know it is more informative to observe a student during a mathematical activity than to grade his papers" (p. 84). Observing students as they work can reveal qualities of their thinking that are not tapped by written or oral activities (National Council of Teachers of Mathematics [NCTM], 1995). When observations are applied continuously in every mathematics lesson, such short-cycle assessments (Wiliam & Leahy, 2015) allow you to determine and monitor the focus of what's being taught as well as continually update student progress.

Observation is about paying attention to what you see in the classroom and spending time anticipating what you might notice or observe. It's a form of **professional noticing**—"a set of interrelated skills which include attending to children's strategies, interpreting children's understandings, and deciding how to respond on the basis of children's understandings" (Jacobs et al., 2010, p. 172).

So, let's get started! As indicated in the previous module, our intent is to engage you in both thinking about and preparing to use the Formative 5 techniques. Complete the What About You? activity that follows. If possible, share your comments with a teacher colleague.

What About You? Observations

1. Think about a lesson you have recently taught. What was the mathematics content focus of your lesson?

____

____

____

____

2. What did you observe? Briefly describe a few (three or four) observation "snapshots" you recall from your lesson.

____

____

____

____

3. What did you notice about these observation snapshots? For example:

a. Which of your observation snapshots were about your students being engaged in the mathematics they were learning? What did you observe?

____

____

____

____

b. Describe something you observed that you actually anticipated. Were you in any way surprised by what you observed?

____

____

____

____

c. Which of the observation snapshots would be described as observing student behavior, rather than observing students engaging in doing mathematics?

____

____

____

____

d. What was the most interesting or surprising thing you observed?

Ponder This: You observe your students all day long. How do these daily observation snapshots influence each of the following?

- Your ongoing assessment of your students
- The feedback you provide your students, and the timeliness of the feedback
- Your planning for the next day

Video—Planning for Observations

Video 1.1

https://bit.ly/3gLKhWZ

Jon Wray introduces the importance of planning for observation in terms of specific mathematical thinking and behaviors you are expecting to observe. Jenny, a seventh-grade mathematics teacher, and Anne, a special education teacher, co-plan their observations for their next lesson. Notice how Jenny and Anne answer each of the four planning questions based on their task of creating windows for the wall of the house. Then respond to the following focus questions:

1. What did you notice about the detail and precision with which Jenny and Anne plan for their observations?

2. How does determining the lesson target influence what they plan to observe?

3. What else do you notice about their conversation as they plan for their observations?

PLANNING FOR OBSERVATIONS

So, let's begin to think about *your* planning for the use of observation as a formative assessment technique. The first step in your daily planning, of course, is determining the mathematical focus of a lesson and how your students will be engaged in the mathematics they are learning. Along with such mathematical and instructional intentions should be the consideration of how you will use observation. Luke's narrative as follows is fairly typical of early connections with planning and the use of formative assessment, generally, and the Observations technique.

> *"I've been teaching for eight years, and I never thought much about formative assessment. It seemed vague—so many ideas about what it was and how I should be using it—but I was never really plugged into why and the actual specifics of how to use it. So, over the last year or so as we have been using what we call the 'Formative 5,' when I sit down after school or at night and actually plan my lesson, I have gotten to the point where I anticipate what I might observe as my lesson is implemented and think hard about what it might look like. Then, pretty important here, I think about what I might do with what I observe—whether that's moving to another of the Formative 5 techniques or considering feedback with my students or thinking about what I have just observed and how it will impact my planning for the next day. This has all become pretty routine now, but I admit that I could have always done this; I just never realized the power and importance of observing."*

Luke's growth in understanding the value of observation and using this technique helped in the development of the guiding questions for daily use of observation and its connection to *your* planning and teaching. The questions in Figure 1.1 are intended to guide you as you consider whether and how you will use observations to monitor instruction and student learning.

FIGURE 1.1 • **Planning for Observations**

1. What will you expect to observe?
• As you plan a lesson, anticipate what you expect students to be doing as they engage in the mathematics focus of the day's lesson.
2. How will you know "it" if you see it?
• As you plan and then think about teaching a lesson, how will you know if what you expect to observe actually occurs? • This consideration sharpens the first question and extends it from what you anticipate or expect to the actual reality of considering responses—and that's assessment.

INSIGHT

Consider both strengths that you might observe and how those observed strengths could be leveraged to advance student learning.

3. What particular strengths or challenges might you observe?
• As you get ready to finalize how and what you will observe within a lesson, it just makes sense to reflect on your experience in teaching mathematics. • What are the "bumps" in your lesson that you may want to especially look for or observe? • Are such "bumps" related to conceptual understandings, procedures, use of representations, use of specific mathematics vocabulary, being able to write a response to a problem's solution, a student's social-emotional readiness for the activity, or something else?
4. How will you record and provide feedback of what you observe?
• What tools will you use, and how frequently will you use them, to more formally monitor what you observe? • In particular, see the "Tools for Using Observations in the Classroom" section of this module for examples of individual student, small-group, and classroom observation tools. • As you observe your students engaged in the mathematics they are learning, will you provide oral feedback to students as they work, or will you use the tools provided in Figures 1.6–1.9 to provide feedback at another time (perhaps at the end of the lesson or even the end of the day)? • Think about how you can provide opportunities for the class or student groups to provide feedback to you (student-to-teacher) and each other (student-to-student).

INSIGHT
Be sure to provide feedback that recognizes what students are doing well and why such "stepping stones" are important to understanding mathematics.

The lesson examples for Grades 1, 5, and 7 and high school algebra, provided in Figures 1.2, 1.3, 1.4, and 1.5, use the Planning for Observations Template to help you connect these guiding questions to your daily planning for using observations. Take a look!

FIGURE 1.2 • **Planning for Observations Template Example for Grade 1**

INSIGHT
Consider how you might leverage what students understand to address challenges and/or levels of understanding.

Lesson Objective: Students will compare a pair of two-digit numbers based on meanings of the tens and ones digits, recording their comparisons using the symbols >, =, and <, and create orally presented story problems involving the comparison of two-digit numbers. Consider the following as you plan such a lesson.	
What will you expect to observe?  *Source:* iStock.com/Elvinagraph	• Students will work together in small groups as they compare two-digit whole numbers. (Note: Google Slides will be randomly presented. Students will respond on work mats.) • Students, working in groups of three, will use handfuls of counters to compare the number of counters in each of two groups. • Students will work individually to compare amounts of counters and also compare numbers. ***Your* Thinking:** What else might you anticipate observing, particularly given *your* class and *your* students?

How will you know "it" if you see it?  *Source:* iStock.com/VectorCookies	You will see and hear students sharing comments about whether a number is greater than, equal to, or less than another number (e.g., 34 is greater than 21). You will see and hear students use the <, =, and > symbols as they compare the two-digit numbers (e.g., 42 > 34). You will hear students create their own story problems involving comparing numbers. ***Your* Thinking:** What other "its" might you see and/or hear?
What particular strengths or challenges might you observe?  *Source:* iStock.com/Brownfalcon	**Strength:** Students successfully use counters and the <, =, and > symbols to compare two-digit numbers. Students create and verbalize story problems involving the comparison of two-digit whole numbers. **Challenge:** Students have difficulty comparing two-digit numbers beyond a certain number (e.g., they're challenged comparing numbers greater than 50 or comparing numbers closer to 100). Students are unsure when stating a comparison and using the symbols (e.g., is it 34 < 40 or 40 > 34?). **Challenge:** Confusion or partial understanding—students seem unable to determine the meaning and use of the <, =, and > symbols. **Strength and Challenge:** Students are more comfortable using counters as they compare numbers. ***Your* Thinking:** What particular strengths or possible challenges have you seen/experienced that may occur?
How will you record and provide feedback of what you observe?  *Source:* iStock.com/Rifai Ozil	Consider the examples of the individual student, small-group, and class observation tools in Figures 1.6–1.9. You can access these tools for your own use at **https://qrs.ly/wsetnnz**. Consider taking a picture of what you observe as a record of student performance. Consider an observed response that may require immediate (typically) oral feedback. Think about how *you* might provide feedback to your students using your responses to the Planning for Observations questions (Figure 1.1). Also, consider opportunities for student-to-teacher and student-to-student feedback.

FIGURE 1.3 • **Planning for Observations Template Example for Grade 5**

<table>
<tr><td colspan="2">Lesson Objective: Students will divide a (nonzero) whole number by a unit fraction (e.g., $3 \div \frac{1}{4}$), demonstrating their understanding by creating a word problem and representing the solution using a visual model. Consider the following as you plan such a lesson.</td></tr>
<tr><td>What will you expect to observe?

Source: iStock.com/Elvinagraph</td><td>• Students will work together in small groups as they create their word problems.
• Students will use the number line or an area model to partition 3 by “number line hops” of $\frac{1}{4}$ (dividing by $\frac{1}{4}$).
• Students will use a drawing to create three objects and then divide each object into fourths.
• Do you anticipate that students will recognize that $3 \div \frac{1}{4} = 12$ since $\frac{1}{4} \times 12 = 3$ (or $12 \times \frac{1}{4} = 3$)?
• Your Thinking: What else might you anticipate observing, particularly given your class and your students?</td></tr>
<tr><td>How will you know “it” if you see it?

Source: iStock.com/VectorCookies</td><td>You will see and hear students sharing word problems for $3 \div \frac{1}{4} = 12$, showing the number of $\frac{1}{4}$s in 3 on their number lines.
In addition to an appropriate word problem, you will see students use manipulatives, an area model, or the number line to represent the problem and solution for $3 \div \frac{1}{4} = 12$ recognizing that $3 \div \frac{1}{4} = 12$ can be thought of as $\frac{1}{4} \times 12 = 3$.
Your Thinking: What other “its” might you see and/or hear?</td></tr>
<tr><td>What particular strengths or challenges might you observe?

Source: iStock.com/Brownfalcon</td><td>Strength: Students successfully use their number lines, drawings, and manipulatives to interpret and solve their word problems.
Challenge: Students have difficulty framing word problem contexts for $3 \div \frac{1}{4} = 12$.
Challenge: Confusion or possible misconception—students seem unable to recognize the relationship between $3 \div \frac{1}{4} = 12$ and $\frac{1}{4} \times 12 = 3$.
Challenge: Students have limited experience using representations for division of whole numbers by unit fraction problems.
Strength and Challenge: Students are more comfortable using area models than the number line for representing division of whole numbers by unit fraction problems.
Your Thinking: What particular strengths or possible areas of challenge have you seen/experienced that may occur?</td></tr>
</table>

How will you record and provide feedback of what you observe?  *Source:* iStock.com/Rifai Ozil	Consider the examples of the individual student, small-group, and class observation tools in Figures 1.6–1.9. You can access these tools for your own use at **https://qrs.ly/wsetnnz**. Consider taking a picture of what you observe as a record of student performance. Consider an observed response that may require immediate (typically) oral feedback. Think about how *you* might provide feedback to your students using your responses to the Planning for Observations questions (Figure 1.1). Also, consider opportunities for student-to-teacher and student-to-student feedback.

FIGURE 1.4 • **Planning for Observations Template Example for Grade 7**

Lesson Objective: Students will test for equivalent ratios by using a ratio table or graphing on a coordinate plane and observing whether the graph is a straight line through the origin. Consider the following as you plan such a lesson.	
What will you expect to observe?  *Source:* iStock.com/Elvinagraph	• You will see that students recognize how earlier understandings related to fraction equivalence are related to determining proportional relationships. • You will see students comfortably using ratio tables to represent and help create equal ratios/proportions. • You will see students using coordinate graphs to represent and then define proportional relationships. ***Your* Thinking:** What else might you anticipate observing, particularly given *your* class and *your* students?
How will you know "it" if you see it?  *Source:* iStock.com/VectorCookies	You will see students using ratio tables to create proportions. Students will accurately describe how they created or validated proportional relationships using the ratio table and/or the coordinate plane. ***Your* Thinking:** What other "its" might you see and/or hear?
What particular strengths or challenges might you observe?  *Source:* iStock.com/Brownfalcon	**Strength:** Students are able to connect prior learning related to equivalent fractions to proportional relationships. **Challenge:** Students are not comfortable in their use of the ratio table or coordinate plane. **Challenge:** Students are unable to create varied representations of proportional relationships. ***Your* Thinking:** What particular strengths, possible areas of confusion, or challenges have you seen/experienced that may occur?

How will you record and provide feedback of what you observe?  *Source:* iStock.com/Rifai Ozil	Consider the examples of the individual student, small-group, and class observation tools in Figures 1.6–1.9. You can access these tools for your own use at **https://qrs.ly/wsetnnz**. Consider taking a picture of what you observe as a record of student performance. Consider an observed response that may require immediate (typically) oral feedback. Think about how *you* might provide feedback to your students using your responses to the Planning for Observations questions (Figure 1.1). Also, consider opportunities for student-to-teacher and student-to-student feedback.

FIGURE 1.5 • **Planning for Observations Template Example for High School Algebra**

Lesson Objective: Students will estimate or calculate the average rate of change over a specified interval.	
What will you expect to observe?  *Source:* iStock.com/Elvinagraph	• Students will recognize how earlier understandings relate to how functions change by describing them as increasing, decreasing, or staying constant. • Students will give attention to units in calculating or estimating average rate of change, contributing to how much the output changes relative to the input quantity. ***Your* Thinking:** What else might you anticipate observing, particularly given *your* class and *your* students?
How will you know "it" if you see it?  *Source:* iStock.com/VectorCookies	You will observe students engaged in discussions revealing that they understand that the average rate of change is a measure of how much the function changes per unit, on average, over an interval. Students will state that if the two points on the graph of the function are (*a, f*(*a*)) and (*b, f*(*b*)), the average rate of change is the slope of the line that connects the two points. You will see students calculating the average rate of change of a function by dividing the difference in the outputs by the difference in the inputs, or $\frac{f(b)-f(a)}{b-a}$ ***Your* Thinking:** What other "its" might you see and/or hear?

What particular strengths or challenges might you observe?  *Source:* iStock.com/Brownfalcon	**Strength:** Students are able to calculate the average rate of change over the specified interval provided. **Challenge:** Students have difficulty visualizing and understanding that finding the average rate of change is equivalent to finding the slope of the line connecting two points. **Challenge:** Students are not comfortable approximating the distribution of points on a scatterplot without worrying about the smaller changes between them. ***Your* Thinking:** What particular strengths, possible areas of confusion or challenges have you seen/experienced that may occur?
How will you record and provide feedback of what you observe?  *Source:* iStock.com/Rifai Ozil	Consider the examples of the individual student, small-group, and class observation tools in Figures 1.6–1.9. You can access these tools for your own use at **https://qrs.ly/wsetnnz**. Consider taking a picture of what you observe as a record of student performance. Consider an observed response that may require immediate (typically) oral feedback. Think about how *you* might provide feedback to your students using your responses to the Planning for Observations questions (Figure 1.1). Also, consider opportunities for student-to-teacher and student-to-student feedback.

INSIGHT

A blank version of the Planning for Observations Template is available for you to download at https://qrs.ly/wsetnnz.

These grade-level and high school examples are all related. They demonstrate how you might use the questions provided in Figure 1.1 as you plan for using observation within a particular lesson. Note that the considerations for an observation will vary with each lesson's mathematics content focus and your students' prior knowledge.

Now let's think about how you might provide feedback to your students. We'll start with an example that is similar to our fifth-grade lesson example (Figure 1.3).

Focusing on Feedback

Imagine that you observed a student writing the following word problem and providing the accompanying drawing:

"Mabel has $\frac{1}{2}$ brownie and wants to share it with 4 friends. How much brownie will each friend get?"

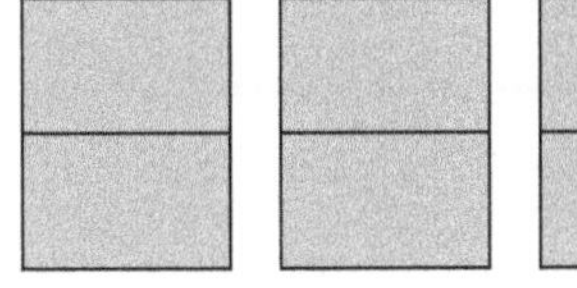

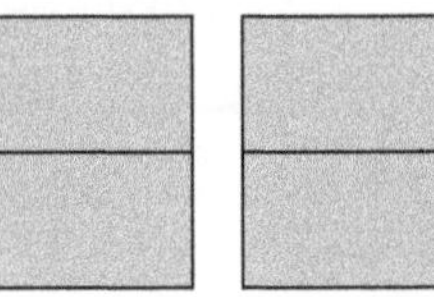

(Continued)

(*Continued*)

- What teacher-to-student feedback would you provide?
- What opportunity for student-to-student feedback could you facilitate?
- Which type of feedback, provided in the grade-level (Grades 1, 5, 7) and high school algebra lesson examples, would you consider for your observations?
- As you observe, when and how do you provide feedback to your students or provide opportunities for your students to provide feedback to you? Share your responses with a colleague.

INSIGHT

Let students know you are observing them as they engage in a task, use manipulatives or other representation tools, and/or work collaboratively to solve problems. Then share what you observed with the students to recognize and support their mathematical thinking.

TOOLS FOR USING OBSERVATIONS IN THE CLASSROOM

The tools provided in this section of the module should assist you as you plan for and regularly use observation as a formative assessment technique. The tools are related to monitoring small-group, whole-class, and individual student observations, respectively.

Small-Group Implementation and Recording Tool for Observations

You can use the Small-Group Implementation and Recording Tool for Observations to make note of what you observe in a small-group instructional setting and record student responses useful for providing teacher-to-student feedback. Figure 1.6 is an actual example of how one teacher used this tool. Some teachers like to carry a copy of this tool around and just jot down comments about what they observe. Others complete it at the end of the day. Either way, this tool serves as a documented record of observations of your students as they work in small groups. You can download this tool at **https://qrs.ly/wsetnnz.**

FIGURE 1.6 • **Small-Group Implementation and Recording Tool for Observations**

INTENT OF THE OBSERVATION	BRIEF DESCRIPTION/ COMMENTS	OBSERVED?
Mathematics Content	Students (six of seven) successfully represented equivalent fractions on the number line; I am not sure about one student's understanding of equivalence.	Yes, for six of the seven students; I need to set up an interview with one student. She seems to be engaged in the lesson; I'm just curious about her comments during the group's discussions.

INTENT OF THE OBSERVATION	BRIEF DESCRIPTION/ COMMENTS	OBSERVED?
Mathematical Practices/Processes	Reasoning, using tools.	Yes, but several of the students seemed just a little unsure about getting started with the number lines.
Student Engagement	Using the number line, seven students were to represent four fractions equivalent to $\frac{1}{2}$ and discuss how they represented the equivalent fractions.	I think I may need to spend a little more time tomorrow discussing their reasoning for placing the fractions on the number line and equivalence, in general.
General Comment: Overall, this group seemed to do pretty well with this activity. I need to work more directly with one student who seems unsure about actual use of the number line and equivalence. I also need to help the group, maybe even my whole class, with just constructing number lines.		
Feedback: To students: I will be interviewing one of the students and then talking briefly to the other six as a group. I quickly checked each of the student number lines and asked for a validation of what they did and why—pleased with their response, particularly for a second lesson involving equivalent fractions. I will ask the group to provide some feedback to me regarding their understanding of this task, what they have learned, and how I might improve this activity.		

Source: Fennell, F., Kobett, B., & Wray, J. (2015). Classroom-based formative assessments: Guiding teaching and learning. In C. Suurtamm (Ed.) & A. McDuffie (Series Ed.), *Annual perspectives in mathematics education: Assessment to enhance teaching and learning* (pp. 51–62). National Council of Teachers of Mathematics. Republished with permission of the National Council of Teachers of Mathematics; permission conveyed through Copyright Clearance Center, Inc.

A blank version of this tool is available for download at **https://qrs.ly/wsetnnz**.

INSIGHT

Note the final Feedback statement, where the teacher wants to provide time for students to provide feedback regarding their understanding of the task, what they learned, and how the teacher might improve the activity. Providing feedback opportunities in this direction (back to you, from student to teacher) is important and a productive use of time.

Classroom Observations Checklist

The Classroom Observations Checklist allows you to keep quick notes for all students individually. It does not provide the opportunity to describe what you have observed in depth like Figure 1.6, but it does allow for a brief comment and a quick way to monitor all of your students. You can use it for one day; a few days, as noted in the example in Figure 1.7; or the entire week. Note that the comments on the example are quick notes. The advantage of this tool is that it's quite efficient to use, and it allows you to observe and record comments about as many of the students in your class as you like. Access this tool from **https://qrs.ly/wsetnnz** and adapt it to your needs.

FIGURE 1.7 • **Classroom Observations Checklist**

Unit: Grade 4—Whole Number Operations and Algebraic Thinking			**Date:** 3/7 to 3/10
Lesson Focus: Three days—Problem solving involving whole numbers; factors and multiples.			
STUDENT NAME	**MATH FOCUS**	**MATH FOCUS**	**MATH FOCUS**
	3/7 Use the four operations with whole numbers to solve problems.	3/9 Gain familiarity with factors and multiples.	3/10 Continued work with factors.
Brett	On task, doing well.	Initially seemed confused since it took her several minutes to get started. But once she began she really seemed to just take off!	She's got this—very pleased with herself.
Yasmin	When multiplying, used the partial products method with base ten blocks. On task, but I noticed that she seemed to be somewhat confused about how to use the base ten blocks.	Seemed to understand multiples as just counting by that number. Determining factors was more of a challenge. Will work on this tomorrow.	Yasmin asked me to help her at least three times during her work on the task. I will spend time with her tomorrow, using an interview.
Matteo	Completed the task fluently, used some mental math; I need to provide a more challenging follow-up example.	Liked this; he's engaged. Today's tasks worked better than the previous lesson.	Will interview and provide a particularly challenging task involving factors and ask Matteo to discuss his solution strategies with me.
Hannah	On task.	On task.	On task.
Alanna	On task.	Slow start, but then seemed to catch up with everyone. Finished early.	On task.

INSIGHT
Notice that this tool uses teacher observation to "check in" with individual students and get a sense of the class's performance across the three days noted. You may not want to use this tool daily. Would it be more valuable for introductory lessons on a particular topic or skill? Would it be helpful to advise your classroom grouping? When might you use this tool, and why?

Hanein	On task, but behavioral management issues within her group.	Knows I'm watching, more focused than earlier. Very successful with multiples and factors activities.	On task; management is no longer an issue.
Advik	Doing well.	Continues to do well.	Has had a great day.
Jordan	Initially seemed confused when solving division examples, in particular. Then seemed to struggle for much of the lesson's activities.	Need to interview and possibly do Show Me with him.	May not get to this today.
Heather	Need to work with individually—ASAP.	Did not start this activity—she's not ready.	Will do this tomorrow.
Kasey	Disrupting progress of her group.	Settled down, worked with a partner, and had a lively discussion about the factors of 24.	Doing fine.
Add more rows as needed to accommodate all members of the class.			

A blank version of this tool is available for download at **https://qrs.ly/wsetnnz.**

Individual Student Mathematics Strengths Observation Log

The Individual Student Mathematics Strengths Observation Log helps you document observations for individual students in greater depth. Note that it accounts for areas of a student's progress, which includes mathematical disposition, how memory may play a part in a student's response, a student's attention to elements of the mathematics experienced, social-emotional elements of the learner, and organizational skills. This tool provides a helpful reminder of the importance of strengths-based teaching. These strengths can and should be leveraged to advance student learning. You might use the tool in the following way:

- Mathematics Concepts/Skills: Record what you have observed mathematics-wise in the lesson or lessons recently observed.
- Mathematical Disposition: Note that the example response in Figure 1.8 indicates how the student approaches the mathematics learning opportunities of the day.

FIGURE 1.8 • **Individual Student: Mathematics Strengths Observation Log**

Mathematics Strengths Observation Log for: *Maria, Grade 2*				Date: *3/14*	
	LEARNER PROFILE				
MATHEMATICS CONCEPTS/ SKILLS	**MATHEMATICAL DISPOSITION**	**MEMORY**	**ATTENTION**	**SOCIOEMOTIONAL**	**ORGANIZATIONAL SKILLS**
List the student strengths with specific content, concepts, and skills.	What types of content, tasks, and activities does the student respond to with positivity, interest, and engagement?	What kinds of things does the student remember?	What strengths does the student demonstrate? Does the student attend to particular types of activities?	How well does the student: • Work with others? • Productively struggle? • Persist?	How does the student organize/ record thinking for mathematics?
Maria does well with activities involving counting and place value.	*Maria seems interested in today's lesson and is generally interested in mathematics and almost always fully engaged. She does seem to get frustrated with problems involving the situations for addition and subtraction.*	*Memory is not a factor in today's lesson, but does seem to slow Maria up on math fact lessons. Maria seems to remember content best when connected to a real-world context.*	*Maria was focused for most of today's lesson, but she seemed to move a little too quickly through one of the lesson's final problems. Maria seems to attend best at the beginning of a lesson and during discussions and group work.*	*Maria was really engaged in the group task today. She led the group in recording responses.*	*Maria needed help getting organized in today's lesson. She organizes her work best when I help her set up a plan before she starts working.*

online resources A blank version of this tool is available for download at **https://qrs.ly/wsetnnz.**

- Memory: Discuss the extent to which the student's memory may have impacted responses. This will be more of a consideration for particular lessons (e.g., vocabulary, basic and related facts, mental mathematics).
- Attention: Consider the student's attention to the activities presented. Note the comment related to Maria's attention.
- Social-Emotional: Address the student's ability and willingness to work productively with others, persist in solving problems, and otherwise engage in the mathematics being presented.
- Organizational: Note how this element of the tool addresses how well the student is organized to attend to the demands of the mathematics activities central to the lesson.

As noted, you would consider this tool for an individual student you are closely observing, particularly a student for whom a more complete picture of progress may be needed. Our experience has been that many teachers like to use this tool in conjunction with or as an interview (Module 2). Some teachers also like to use the tool to collect observed student strengths over time.

INSIGHT

Notice that the tool shown in Figure 1.8 attends to specific strengths and individual needs of the learner. When might you use this tool? How helpful might it be as you organize for parent–teacher conferences?

Individual Student Observation Check-In

The Individual Student Observation Check-In is a tool that provides a quick "check-in" for individual students as seen in Figure 1.9. Often used on a tablet device or 3 × 5 index card, teachers can move around the classroom and just circle the extent to which individual students or a small group of students are productively engaged in the lesson at various stages of the lesson: early, midway, and at the end of the lesson. You can indicate if you would like to extend the observation to include an interview and provide comments related to what you have observed. This functional, easy-to-use tool indicates the extent to which your students are engaged in what you have planned, which is important for your planning for the next day and beyond.

FIGURE 1.9 • **Individual Student Observation Check-In**

Name: Bryce	**Date:** March 14	
Mathematics Focus of the Lesson: Subtracting mixed numbers with like denominators.		
ELEMENTS OF THE LESSON (EARLY, MID, END)	**PRODUCTIVELY ENGAGED**	
Early: Replacing each mixed number with an equivalent fraction.	(Yes)	No
Mid-Lesson: Representing each fraction on a double number line and comparing the differences.	(Yes)	No

(Continued)

(Continued)

End of the Lesson: Move to symbolically subtracting the mixed numbers and discussing the differences.	Yes	(No)
Need for an interview?	(Yes)	No
Comments: Bryce was engaged for most of the lesson and seemed to do well particularly when comparing the equivalent fractions using double number lines. But when we got to just subtracting the mixed numbers, she couldn't connect with what we had done in the first part of the lesson. I will see how she does in tomorrow's lesson, which will continue with subtracting mixed numbers with like denominators. I think I will do a brief interview with Bryce toward the end of tomorrow's lesson.		

A blank version of this tool is available for download at **https://qrs.ly/wsetnnz.**

INSIGHT

Recognize that this tool provides a quick way to monitor individual student progress at particular junctures within a lesson. Also note that paraprofessionals and student interns may be able to use this tool while you are working with a student group.

Consider the following questions and classroom-based responses as helpful suggestions for how you might use the tools presented in Figures 1.6–1.9.

1. **Think About:** I'm not expected to use all of these, all the time, right? I'm not even sure where to start!

Classroom Response: *When I first began thinking hard about how to not just observe what my students are doing in mathematics class each day but also connect to my planning and begin to record what I have observed, the Planning for Observations Template (see Figures 1.2–1.5) became my new best friend! The questions provided on the template—what I then called my planning pal—just made sense, and use of the template really helped me get started in thinking about how to truly consider how and even when to use observation as an assessment technique.*

2. **Think About:** Which of these tools should I use, and when?

Classroom Response: *I had never recorded much of anything I observed my students doing in mathematics class! Wow—has the study and use of the Formative 5 changed my view. I was always grasping for comments at parent–teacher conferences or when I received calls from parents, and I seemingly never had actual "evidence" of what my students were doing or had done. The more I got into seriously reviewing and actually trying out the tools provided for observation, the more I realized that they were organized as small-group (Figure 1.6), whole-class (Figure 1.7), or individual student (Figures 1.8 and 1.9) observation tools. I started with the small-group tool (Figure 1.6) since my independent workstations were organized by small groups. That tool worked well for me. I have also discovered that the tools are very valuable for parent conferences. They provide me with explicit data over time, which helps me explain student mathematical understandings to parents. They love it!*

3. **Think About:** How much of what I observe should I record, and when?

Classroom Response: *I have found that once I considered what I would observe as I planned my lesson, it was easier to determine which of the observation tools I would use, and that influenced how much I would record. In general, I would take notes on a tablet as I moved around the room. Sometimes I used the Classroom Observations Checklist (Figure 1.7) for a full week, or at a minimum several days, so I could see if there were tendencies being exhibited by particular students—having particular mathematical challenges, infrequent engagement in lesson activities, and so on. Sometimes I would add more to my comments after mathematics class or at the end of the day. If I had particular concerns about what a student was doing, it almost always drove me to use one of the two individual student observation tools. Figure 1.8, the Individual Student Mathematics Strengths Observation Log, really examines a variety of learning dimensions that have always been helpful for me to document before I consider an interview for a student, and it requires more documentation than Figure 1.9, the Individual Student Observation Check-In, which is a quick-response tool that essentially monitors student engagement at key points throughout the lesson. Not always, but often, what I have observed using the small-group tool (Figure 1.6) or classroom tool (Figure 1.7) suggested the use of one of the two individual observation tools (Figures 1.8 and 1.9). After almost a year of intentionally using observation as a valued formative assessment technique, I know I am much more comfortable in deciding which observation tools to use, and when and how much to record, all of which provides me with suggestions for the feedback I can give to my students and particular next steps, which may be an interview.*

Video—Conducting Observations

Video 1.2

http://bit.ly/3UpId4C

Co-teachers Jenny (mathematics teacher) and Anne (special education teacher) and their mathematics coach, Greta, discuss the use of observation and its connection to feedback.

Think about and discuss how Jenny and Anne used observation in their co-taught lesson and how they connected their observations to feedback. How do you provide feedback opportunities related to what you observe?

Time Out

Let's Reflect:

- Of the observation tools presented and discussed, which seem to be particularly helpful to you? Why?

____

____

____

____

- Which of the tools might you use most frequently?

____

____

____

____

- How might you adapt one or more of the tools for specific planning and instructional needs in *your* classroom(s)?

____

____

____

____

Ponder This:

- Which of the tools may be helpful to you as you plan for parent–teacher conferences? Name the tool and why you think it would be helpful to you when communicating the progress of your students to parents/guardians.

____

____

____

____

INSIGHT

As you use observation as a formative assessment technique, how should feedback (teacher-to-student, student-to-teacher, and student-to-student) be connected to the actual use of observations?

- Having just read a classroom response to the following question—*How much of what I observe should I actually record, and when?*—how would *you* respond to the question?

TECHNOLOGY TIPS AND TOOLS FOR RECORDING OBSERVATIONS

In addition to the tools just presented, there are well-known digital tools that provide teachers with the means to capture student observations, each of which has specific advantages. As with any digital tool that is used to collect student data, be sure to investigate and follow school district data privacy policies and practices, and communicate the privacy plan and purpose of your recordings with parents/guardians and students.

- Every smartphone, computer, or tablet device comes equipped with a digital image and video camera. The images and videos collected during your observations of students can allow both you and your students to document and review student engagement with various mathematical practices or processes, use of representations, and mathematical arguments and reasoning, as well as student preconceptions and challenges—all evidence of student learning. The digital files can be easily captured, stored, and used as powerful formative artifacts for students and teachers, and inform plans for next steps. Having opportunities to pause, analyze, and reexperience an observation captured digitally, rather than relying on memory or written notes, can be extremely beneficial.
- Google Forms (www.google.com/forms) can be used to create your own classroom observation look-fors. This free and easy-to-use resource can be an effective and efficient way of developing your look-fors to capture observation data and seamlessly access your data/notes in a summary or spreadsheet for later analysis.

USING OBSERVATIONS IN *YOUR* CLASSROOM

When thinking about the use of observations to guide and monitor planning and instruction in *your* classroom, consider the following:

When should I do this?

First, and importantly, expect to observe every day to inform your teaching and guide your planning using the tools provided in this module, as appropriate.

What should my observations focus on?

Your focus will be on observing mathematical understandings and student engagement with particular mathematical practices and processes (e.g., reasoning, problem solving, precision, modeling with mathematics). While you will, no doubt, observe off-task or distracting conduct/behavior as your students engage in a mathematics activity, try not to be overly distracted by such student behaviors, which are not the focus of your assessment. But do take the time to note these behaviors as they accumulate, and attend to them as needed, as it will be important to consider if and how such behavior may be impacting a student's level of engagement and performance in mathematics. Observing, or perhaps even informally timing, how long students stay on task while completing a mathematics problem or related assignments may be one way to judge both the level of student engagement in learning mathematics and the possible impact of student behavior on mathematics performance. Some teachers make a note on the Small-Group Implementation and Recording Tool for Observations (Figure 1.6) or the Classroom Observations Checklist (Figure 1.7), noting when students are particularly engaged and how frequently off-task and distracting behaviors are occurring. We also note that students who don't understand and are unsure of expectations may be more likely to be off task, however defined, than others.

How should I use the tools to support my planning and observing?

The more you become used to anticipating and then planning for how and what you will observe within a lesson, the more you can focus on really watching and listening to your students during a lesson. The Observations technique and the tools provided in this module will become an integral component of your planning and teaching. Your observations will range from relatively quick and informal observations to focused, deliberate observations of individual students, small groups, or the entire class. You can mix and adapt the module's tools in any way that works best for you and your students. As you anticipate what you will observe, your observations will become a natural way of noticing what students are doing mathematically; you will then begin to seamlessly use what you have observed to guide and monitor student and class performance. Keep the following in mind as you plan for and regularly use observations as a classroom-based formative assessment technique:

> **INSIGHT**
> *Consider observing and noting the kinds of tasks and problems that generate enthusiasm and excitement among students.*

- Remember that observation should be intentionally connected to the actual planning and implementation of the day's lesson. This is a key point.
- This is *your* classroom. Expect a full range of responses. Recognize when students are challenged, when they seemingly "sail through" an activity, or when they exhibit signs of confusion or frustration.

- *Principles to Actions* (NCTM, 2014) notes that effective teaching of mathematics engages students in solving and discussing tasks that promote mathematical reasoning and problem solving. How can observations be used to identify student strengths, as instructional starting points, as well as validate the level of engagement you desire? Note that the Individual Student Observation Check-In (Figure 1.9) addresses the extent to which students are engaged at key points throughout a lesson.
- Document, document, document. Keeping a record and analysis of specifically what you observe (e.g., particular use of varied representations) will more directly inform decisions during the lesson's implementation and advise your short-term and long-term planning. As importantly, such records and reflective comments can be used as evidence of growth and be shared with students and parents or caregivers to celebrate milestones.

Time to Try

As noted earlier, you observe your students all day long. And what you observe provides a continuing update of their strengths, understandings, instructional setbacks, and challenges. The Observations module provides the rationale and validation for the importance of observation as a classroom-based formative assessment technique. So, let's try this out! Consider a mathematics lesson you may plan for and implement tomorrow. Yes, tomorrow! Use the following adaptation of the Planning for Observations Template (Figures 1.2–1.5) to enter in brief responses to each of the concerns related to planning for and using observation as you teach. Discuss your "plan" with a colleague.

Planning for Observations Template (Adapted)

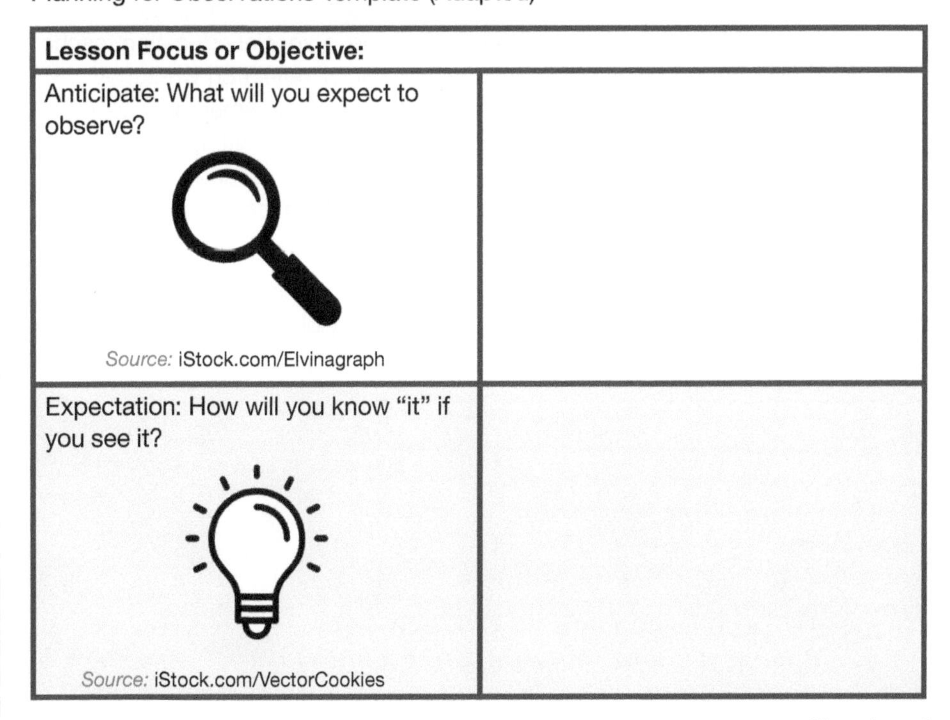

Lesson Focus or Objective:	
Anticipate: What will you expect to observe? *Source:* iStock.com/Elvinagraph	
Expectation: How will you know "it" if you see it? *Source:* iStock.com/VectorCookies	

(Continued)

(Continued)

Tools: What tools will you use (see Figures 1.6–1.9)?	
What particular strengths or challenges might you observe?  *Source:* iStock.com/Brownfalcon	
How will you record and provide feedback of what you observe?  *Source:* iStock.com/Rifai Ozil	

Video—Reflecting on Observations

Video 1.3
http://bit.ly/3Fic4HF

Greta, a mathematics coach; Anne, a special education teacher; and Jenny, a mathematics teacher, all reflect on what they observed within their co-planned and -implemented lesson. Comments include the importance of anticipating what they might observe, the feedback provided to students, and how they will be able to use particular student challenges they observe to plan for the next day's lesson.

Think about and discuss how anticipating what they might observe impacted their co-planning, and how what they observed as the lesson was implemented may impact their planning for the next day.

SUMMING UP

As a classroom-based formative assessment technique, observation provides that initial and ongoing link to planning and instruction. It allows you to consider, before the lesson is taught, what students will do, how they might engage in the mathematics, possible lesson products, and issues related to student grouping, differentiation, and much more. The potential of this preteaching "look for" or noticing opportunity should influence your planning—every day. Additionally, what you observe within a lesson should be the catalyst for the next lesson's planning and instruction and will provide day-to-day anecdotal indicators of student progress and help you make the most of your lesson. Anticipating what you will observe will help you to determine lesson activities, problem-based tasks, and questions. Utilize the Planning for Observations Template (Figures 1.2–1.5) and the observation tools (Figures 1.6–1.9) presented and discussed in this module. Records of what you observe will provide a pattern of student performance that is useful for monitoring progress, providing feedback to students, or guiding conferences with parents/family and others. Keeping a record of observations should also influence the pace of your lessons and decision making within a lesson, as well as provide indicators for additional longer-term planning and instruction. Finally, your use of the Observations technique on a regular basis will identify the need for the regular use of Interviews, the second element of the Formative 5 palette of classroom-based formative assessment techniques.

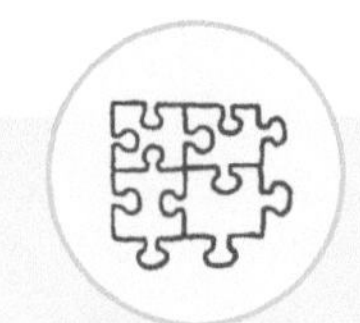

Your Turn

Rate, Read, Reflect! Consider the following questions, then rate (items 1–3) and discuss your responses (items 4–7) with your grade-level or mathematics department teaching team or with teams across multiple grade levels or mathematics courses.

1. How important are your observations of students as a formative assessment technique?

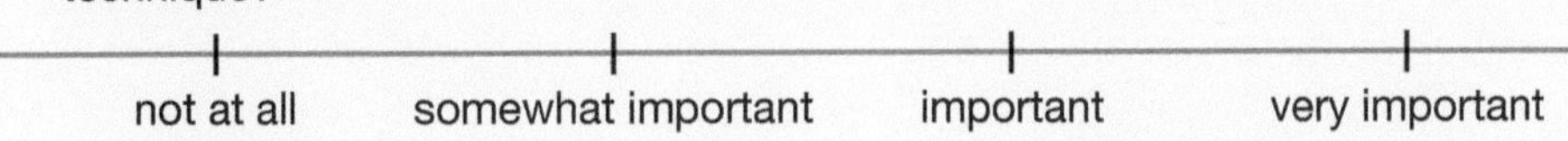

2. How often do you actually use observation to assess student progress in mathematics?

(Continued)

(Continued)

3. When do you make notes about what you have observed in your mathematics classroom?

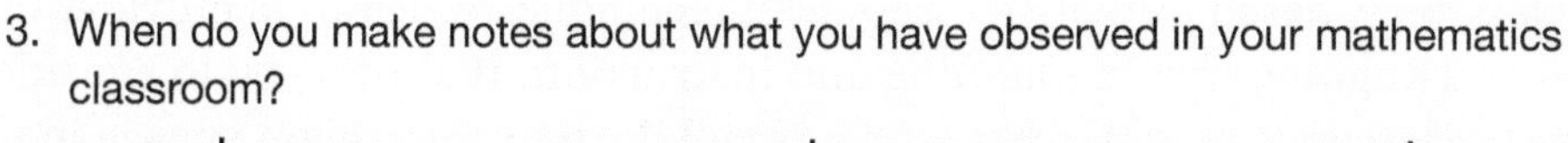

4. As you anticipate *what* you will observe in a mathematics lesson, how will you actually "do" your observing? And *what* will you use to keep a record of what's observed?
5. Which of the observation tools in this module might you use or adapt and use in your own classroom? What observation tools have you created or used?
6. When observing students doing mathematics, *what* is the feedback that you provide? How do you provide opportunities for your students to provide feedback to you, and what about opportunities for students to provide feedback to each other?
7. Professional Connection: Consider reading *Professional Noticing of Children's Mathematical Thinking* by Victoria R. Jacobs and colleagues (2010), available from http://www.sci.sdsu.edu/CRMSE/msed/papers/Lamb1.pdf, then discuss the following:
   a. How would you describe connections between professional noticing and observations?
   b. Why should observations and their use be an important element of preservice teacher educator learning related to assessment, both formative and summative?

NOTES

MODULE 2

INTERVIEWS

"I observe and talk with my students all day long, every single day. Now I know how to formalize my comments and conduct an interview. So powerful!"

—KINDERGARTEN TEACHER

"For some reason I thought that you only interviewed students who were having problems in math class. Now I regularly interview my algebra students because I want to assess how they are transitioning to using equations and inequalities in a more formal way."

—EIGHTH-GRADE TEACHER

"I have recently started to use small-group interviews to essentially 'check in' with my students. I often ask them to share their thinking about a student's response to a mathematics task. It's so interesting to just listen to how they provide feedback to each other."

—HIGH SCHOOL MATHEMATICS TEACHER

FROM THE CLASSROOM

Jeremy and I were in an after-school professional learning session and discussing formative assessment. Both my elementary school and Jeremy's middle school faculty were involved. We were discussing when you might use an interview with a student or group of students. He teaches middle school math, and I teach first grade. Our instructor was noting the difference between a clinical (diagnostic) interview and the interviews you might use to guide classroom instruction. I pointed out that I used interviews pretty much every day. I have a corner in my room, which I call the Chat Corner, and I regularly have a student meet with me just to "check in" on something I may have observed. My interviews are quick—a couple of minutes. I usually just point to a problem and ask my student "How did you do that?" and then "Why did you do it that way?" Jeremy was blown away. He exclaimed, "Wow, I have never done that! But I could and probably should. Seems so simple and so valuable." Two weeks later, our groups came together again, and Jeremy thanked me for my "interview story." He had created his own recording tool to help him see who he had interviewed and why, and then also recorded what he observed the next day or so. We agreed that the few minutes we both now spend doing our own version of the interview are amazingly valuable.

Purpose

There are levels of the classroom-based interview. They may be informal or perhaps follow a more formalized structure. An interview is the next natural step beyond the completion of an observation. It is an opportunity to ask questions to gain further clarification and to reflect on what occurred. It, essentially, allows you to "dig deeper" regarding student thinking. In this module you will not only learn about how to plan for the interview but also learn about some of the most efficient and effective tools for using interviews in the classroom.

Module Goals

As you read and complete activities within this module, you will:

- ✓ Consider the importance of the interview for particular mathematics topics.
- ✓ Reflect on ways to provide feedback to students and/or parents/families using interviews.
- ✓ Plan, use, and reflect on the role of one or more of the interview tools provided in this module.
- ✓ Develop a plan for incorporating interviews into your everyday teaching and related use of classroom-based formative assessment.

INTERVIEWS: BACKGROUND AND BASICS

The practice of using interviews as an assessment technique has a long history and includes connections to mathematics, special education, and feedback (Fennell, 1998, 2011; Ginsburg, 1997; Leighton, 2019; Weaver, 1955). A brief one-on-one or small-group interview has the potential to provide valuable cues regarding student thinking, including reasoning and conceptual understandings. Such cues will identify and guide instructional next steps. Years ago, Weaver (1955) noted that interviewing children allows teachers to study levels of student thinking as they respond to a variety of quantitative situations. Ginsburg (1997) suggests that the use of a clinical interview can follow what a teacher initially observes regarding a written or oral response. Such observed responses are then followed up with questions to engage the student in discussing the "how and why" of their solution strategies.

As an important component of the Formative 5, the interview can be thought of as an informal conversation between teacher and student, or perhaps among a small group of students. Interview questions may be as informal as "How did you do that?" or "Why did you do it that way?" or "Can you explain how you solved that?" (Ginsburg & Dolan, 2011). The use of the interview assumes that rapport between the teacher and student has been established. It's important that you conduct interviews and related conversations in a way that identifies and builds on students' strengths. Think about it: Interviews are conversations, and they should be planned to identify strengths and commit to advocating for student success. The interview should both inform and motivate you and your student(s) (Chappuis et al., 2017).

INSIGHT
As conducted, interview questions encourage students to respond, but do not suggest a particular or expected response.

Interviews are extensions of observations. They are brief and informal conversations between the teacher and an individual or small group that provide a glimpse into student thinking.

For reasons mostly related to instructional time, the interview is usually brief, typically five minutes or less, sometimes up to ten minutes for a more in-depth interview or perhaps an interview that involves a small group of students. Consider the following example of how what's observed can suggest the need for an interview.

The day's lesson involved partitioning regarding fractions and mixed numbers. The students were asked to use materials or drawings to show how they could share 11 cookies among 4 students, and $\frac{1}{2}$ of a large cake among 3 students. Janet's response is provided in Figures 2.1 and 2.2. What Janet did was quite interesting as her representations differ for each of the problems. For sharing 11 cookies with 4 students, she circled 2 cookies four times, so everybody got at least 2 cookies. Then she divided 2 of the remaining cookies in half and gave a half cookie to each of the 4 students—so each now had $2\frac{1}{2}$ cookies. Finally, she divided the last cookie into fourths, giving an additional $\frac{1}{4}$ cookie to each of the 4 students. So, each student now had $2\frac{3}{4}$ cookies.

When Janet used a drawing to show how she shared $\frac{1}{2}$ of the large cake among 3 students, she drew a rectangular cake and then divided the cake into thirds, indicating that each student would get $\frac{1}{3}$ of the cake.

FIGURE 2.1 • **Janet's Solution—Sharing 11 Cookies With 4 Students**

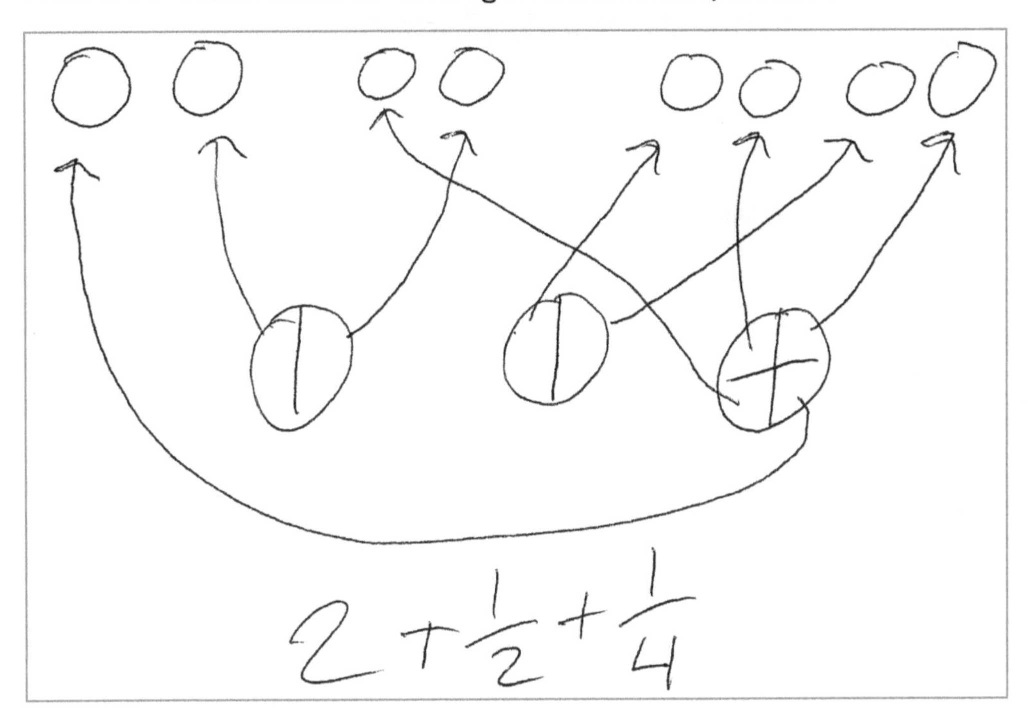

FIGURE 2.2 • **Janet's Solution: Sharing $\frac{1}{3}$ of a Cake Among 3 Students**

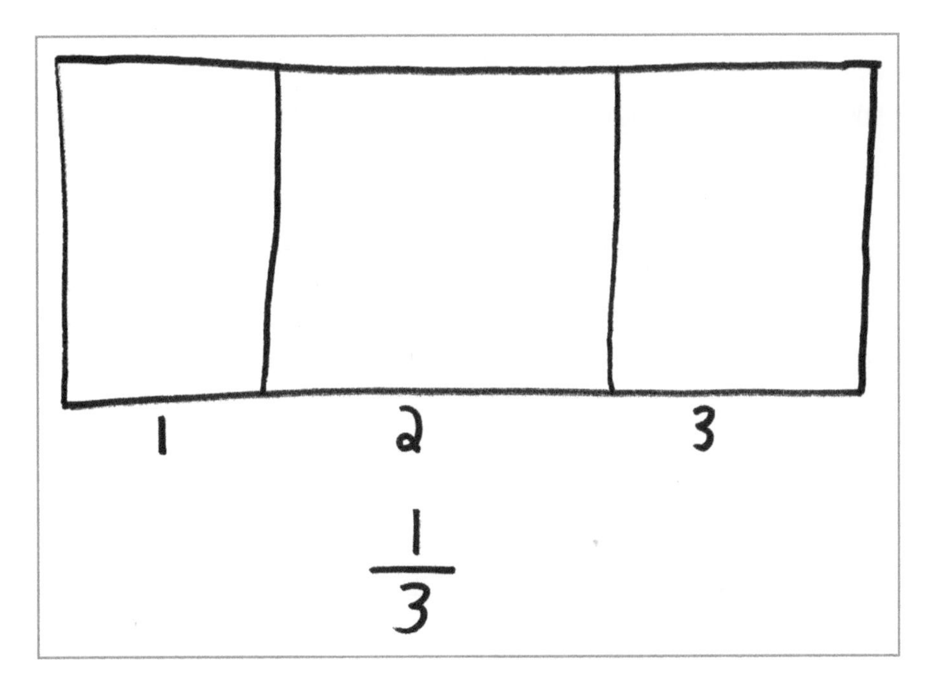

Given Janet's response to the lesson's tasks, would you want to discuss what she did in an interview during the lesson or perhaps some other time that day? What questions might you consider asking Janet? Representations like Janet's are why you might use an interview. Such responses provide you with an opportunity to explore student thinking and understanding and may also be used to jump-start planning and related activities for tomorrow's mathematics lesson.

What About You? Interviews

- Think of a student's response to a mathematics activity that you observed and then decided to talk to the student about. How did such use of an "interview" deepen your understanding of what you observed?

____

____

____

- React to the following statement from a teacher learning about the use of interviews:

  "I am going to try using the interview as a formative assessment in place of some of the paper tasks provided. I feel this will be of much greater value than a paper task, as it gives the opportunity to ***discuss with the students their knowledge and understanding of new concepts, as well as those already taught.***"

____

____

____

- How have you conducted interviews with your students?

____

____

____

- How might you use the interview as a formative assessment technique? And how might you connect your use of observations to using interviews?

____

____

____

- How might you provide teacher-to-student feedback after the interview and allow for student-to-teacher feedback?

____

____

____

- **Video:** Plan for and video-record an interview with one student or a small group of students. Play your video back. Write down two reactions to this interview, and share them with a grade-level or mathematics department colleague.

____

____

____

____

Ponder This:

- What do you think you might do differently as you plan for connecting your use of observations to the use of interviews as a formative assessment technique?

____

____

____

____

PLANNING FOR THE INTERVIEW

We feel very strongly about the importance of the Interviews technique as one of the Formative 5 because it provides the obvious follow-up to an observation or observations you may make when implementing a lesson. This classroom-based interview, which is an adaptation of a more clinical interview, suggests that you spend a few valuable minutes digging deeper with an individual student or perhaps a small group of students. While some teachers immediately think about the practical issues regarding how and when to find time to conduct interviews, our experience confirms how flexible teachers can be once they recognize the utility and impact of the interview. Consider the following comments:

> "At first, I had no idea when I would be able to actually do the interview during the hubbub and buzz of my math time. Then I just created this 'you and me' spot in the corner of the room, and when I observed something I wanted to ask about, I just said, 'It's you and me time, right now.' The kids really seem to look forward to it. It's kind of like their private math time with me! I tried to make the interviews about five minutes or so long and took brief notes on what was discussed. Valuable technique—use it every day!"
>
> **—FIRST-GRADE TEACHER**

"I loved the potential of interviews, but I worried about when I might actually conduct them, and I did not want to wait until the end of the day. So, I use small-group interviews—right around my work table. I do this every day usually based on responses I observe that are pretty similar, but not always. The group sharing of my questions works well for me, and it really helps me plan for individual and class needs."

—THIRD-GRADE TEACHER

"I had never even considered using an interview as a type of formative assessment. As I learned about the potential of a brief interview, sort of extending what I observed, I tried it. Now, as I observe students when I get them involved in newer concepts in my algebra, geometry, and precalculus classes, I use the Individual Student Interview Prompt (Figure 2.6) either during my class or during transition time between my classes. My interviews are short, probably two to three minutes, but particularly helpful in determining what my students can do and helping me think about topics that may need more time instructionally."

— HIGH SCHOOL MATHEMATICS TEACHER

INSIGHT
During the interview, listen to the student or small-group responses. In the interview process, such "listening" includes attending to the extent to which students are engaged in the activity, analyzing nonverbal cues, gauging their comfort in using representations, and other aspects of noticing (Jacobs et al., 2010).

One thing that we have found in our continuing work with the Interviews technique is the need to emphasize that use of the interview is not based on a deficit model. While you may want to interview a student based on a series of common mathematical errors, confusion, or related mathematical learning challenges you observe, you may also want to interview a student because of a particularly unique or advanced response, or perhaps a response you just don't understand—and we have all seen our share! But why do this? The interview identifies insights about student thinking that will inform your planning and instruction, providing you the opportunity to adjust your planning and instruction to address the strengths and needs of your students more appropriately.

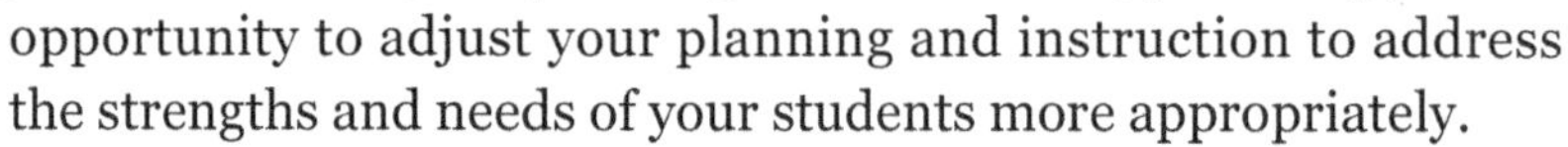

Video 2.1
Jake's 3 + 4 = 8 Interview

http://bit
.ly/3OR8tU8

Video 2.2
Jake's
21 + 23 = 100
Interview

http://bit
.ly/3OR8tU8

Consider the interview with Jake. Jake was somewhat new to the class and had missed a few days, so his teacher started out with the equation 3 + 4 = 8 and asked if it was true or false. Listen to Jake's response by scanning the QR code for Video 2.1.

Next, listen to Jake's responses to 21 + 23 = 100 and 35 + 49 = 100 by scanning the next two QR codes (Videos 2.2 and 2.3).

Note that in all three examples, Jake was just asked to indicate if the number sentence was true or false, but the responses yielded much more about what Jake knew and his thinking. The comments Jake made provide insights that his teacher could use as she planned her next lesson and as she planned for Jake.

As stated in the introductory module, your planning should include everyday attention to the Formative 5 techniques. As you consider the use of the interview, whether with one student or perhaps a small group of students, planning becomes an essential consideration. In this case, your planning will consider how what you expect to actually observe in the next day's lesson may warrant the use of an interview. Based on what you observe (see Module 1), you can use the questions in Figure 2.3 to initially consider and then plan for an interview, keeping in mind, as noted, that such a classroom-focused interview will be relatively brief. Interview questions may relate to any of the following:

Video 2.3
Jake's
35 + 49 = 100
Interview

http://bit.ly/3B1oAbY

- A student's level of understanding of a mathematical concept
- Learning more about a particular strategy being used by one or more of your students
- Extending student thinking
- Assessing comprehension of a task to be solved

Depending on a student's interview response, your follow-up questions can center on a variety of student needs, including digging deeper into a student's reasoning or perhaps extending a student's response by taking it in another direction (e.g., "You stated that 75 was your solution to the equation. Is your response reasonable? How do you know?").

FIGURE 2.3 • **Planning for Interviews**

1. Deciding on an Interview
• Who would you want to interview, and why?
• Would you want to interview students one-on-one or in a small group?
• What might you observe your students doing that would prompt the use of an interview?
• What will you consider when selecting students to interview related to their progress in a lesson? Would you think of a "sampling" of your students—that is, a student who may struggle with the day's lesson, one known for responses you never seem to expect or anticipate, and perhaps a student who you may want to challenge? Or perhaps you would just like to use the interview to check in with one or more of your students, to see how they are doing with the day's mathematical experiences.
2. Setting Up Interviews
• Would you plan to complete the interview during the time you teach mathematics or sometime before or after the lesson?
• Where in the classroom would you conduct the interview?
• For particular areas of concern or interest, should you record the interview using video or audio, or perhaps use an interactive whiteboard app?
• How might you use physical and/or digital representation tools (e.g., algebra tiles, base ten blocks, bar models, pattern blocks, number lines, and drawings) in the interview?

(Continued)

(*Continued*)

3. Interview Questions

- What questions might you ask?
- How might the questions differ for particular students?
- Would you ask students to discuss the "how and why" of what they did within a lesson activity?
- How could you personalize the interview? How could you pose questions so that students feel they are participating in a friendly, thoughtful, respectful, and helpful conversation?

4. Anticipating Interview Responses

- What responses might you anticipate from students? Would such responses reveal conceptual understandings? Would they be related to procedures? Would they involve reasoning and sense making?
- To what extent do you think interview responses would reveal student mathematical dispositions?

5. Planning Follow-Up Questions

- Consider your anticipated interview responses. What follow-up questions to particular student responses might you ask?
- Would such follow-up questions come at the end of the interview or later in the day?
- How might responses to the interview's follow-up questions, and your feedback, influence your planning for the next day's lesson?

Focusing on Feedback

As you plan to interview your students, recognize that once you ask students to explain what they are doing or have done, their response provides you with *student-to-teacher* feedback about the mathematics they are learning. You can expand on that feedback by also asking questions about the context of tasks students are working on (e.g., using mountain range heights as a context for solving problems) or dispositional statements about mathematics, generally (e.g., "Are you enjoying the geometry unit? Why or why not?"). You will also need to think about how and when you will provide *teacher-to-student feedback* based on their interview responses and, importantly, think about how their responses will influence your next planning and instructional steps.

Timing and delivery of feedback during an interview should be strategic and thoughtfully connected to the purpose of the interview. Do you need to gather information about a particular student's understanding to provide just-in-time feedback? Or are you collecting information about student understanding to inform your next instructional steps? Maybe both! Once you have established the intent of your interview, consider the multiple solution pathways that students might demonstrate and your feedback.

Activity: For each scenario that follows, consider the timing of your feedback, type of feedback, and instructional response. Discuss your responses with your grade- or department-level colleagues.

Scenario 1: Students are working on a problem-solving task in pairs. As you walk around the classroom, you notice that one pair of students is engaged in a lively discussion but has not written anything down. You ask, "Tell me about your mathematical ideas." The students begin sharing multiple strategies they are considering. Several of their ideas are strong.

Choose A, B, or C and defend your choices related to the timing and type of feedback and your instructional response.

Timing of Feedback

A. Provide immediate feedback on the strategies the students proposed.

B. Withhold feedback on the strategies until later in the lesson.

C. Provide immediate feedback on the collaborative problem-solving behaviors.

Type of Feedback

A. Direct: Suggest that students adapt or use a particular strategy that they shared.

B. Goal Oriented: Remind students of the goal of the lesson and suggest they choose a solution pathway that connects to the goal.

C. Open: Tell the students that you are interested in hearing more about the solution they choose.

Instructional Response

A. Continue your lesson as planned.

B. Stop the lesson and provide additional guidance to the students.

C. Change direction in the lesson by posing a new question or asking students to share.

Scenario 2: Students are working on a problem-solving task in small groups. You decide to conduct interviews with one individual student "representative" from each of the three groups. You ask, "What strategy are you thinking about using?" and "Have you discussed this idea with your group?" Two students discussed a strategy that was not clear or applied to the task, and one student had a clear solution pathway identified. None of the three students have shared or discussed their thoughts with their groups.

(Continued)

(Continued)

Choose A, B, or C and defend your choices related to the timing and type of feedback and your instructional response.

Timing of Feedback

A. Provide immediate feedback on the strategies the students proposed.

B. Withhold feedback on the strategies until later in the lesson.

C. Provide immediate feedback to the whole group to engage the students interviewed.

Type of Feedback

A. Direct: Suggest that students adapt or use a particular strategy that they shared.

B. Goal Oriented: Remind students of the goal of the lesson and suggest they choose a solution pathway that connects to the goal.

C. Open: Tell the students that you are interested in hearing more about the solution they choose.

Instructional Response

A. Continue your lesson as planned.

B. Stop the lesson and ask the students to engage all group members in the discussion.

C. Stop the lesson and ask some students to share their strategies to jump-start students who could benefit from assistance.

Video—Planning for Interviews

Video 2.4

http://bit.ly/3B0ys5U

Beth Kobett introduces the Interviews technique. Michele, a third- and fourth-grade mathematics teacher, and elementary mathematics specialist Kristen co-plan a lesson discussing how Michele may adapt the lesson based on possible student challenges and her use of the Classroom Interview Record (Figure 2.5). Michele also discusses how she has learned to anticipate how students may respond, and how such anticipation has impacted her planning for and use of the Interviews technique. **Think about and discuss** the role of anticipation in planning for interviews with your own students.

Time Out

Let's Reflect:

- As you plan a lesson, what aspects of your proposed lesson may suggest that you prepare for interviewing individual or small groups of students? Such concerns may, of course, be about the mathematics focus of the lesson, but may also relate to how your students are using representations (e.g., counters, base ten blocks, number lines, drawings, coordinate grids, and graphing technology) or how they are to respond to a lesson's task.

- Consider the following comment from a teacher learning about the use of the interview as a formative assessment technique: *"I'm going to set up an area of my room where I can just suggest that students meet with me for an individual or small-group interview—each day for a quick check-in on how they are doing. I think if I work it into my daily routine, it will become natural to myself and to my students."* What do you think? Is this something you may consider?

(Continued)

(Continued)

- What about connecting interviewing and feedback? You may want to consider having students interview each other relative to a mathematics activity or task they are working on or have completed. The Individual Student Interview Prompt (Figure 2.6) could easily be adapted for this interviewing and student-to-student feedback opportunity. Your students could then also reflect on how their solution pathways were similar or different.

- How could you use responses to your interview questions in planning your mathematics teaching—the next day? Or long term?

- **Video:** As you viewed Michele use the Classroom Interview Record (Figure 2.5) when observing and then interviewing, how was she actually using the tool?

Ponder This:

- How do you think you will find the time to actually do the student or small-group interview?

- As you strive to keep your individual or small-group interviews short (five to ten minutes), consider including a question to prompt student-to-teacher feedback relative to the actual interview process (e.g., "What did you learn during this activity? Would you change it in any way?") or student-to-student feedback (e.g., "Can you share with your partner how their solution strategy helped you to think about the problem differently?").

TOOLS FOR USING INTERVIEWS IN THE CLASSROOM

This module presents four tools for you to consider as you plan for and use the Interviews technique:

- Planning for Interviews Tool (Figure 2.4)
- Classroom Interview Record (Figure 2.5)
- Individual Student Interview Prompt (Figure 2.6)
- Classroom Observations–Interviews–Student Representations Tool (Figure 2.7)

Examples of the tools and comments by teachers who have used them (Think Abouts and Classroom Responses) follow the discussion of each tool. In addition, we suggest online tools that may be helpful to you as you engage in everyday use of the Interviews technique.

The Planning for Interviews Tool

INSIGHT
The Planning for Interviews Tool (Figure 2.4) is a useful guide to help plan for an interview as well as provide a record of student responses.

The Planning for Interviews Tool (Figure 2.4) is a useful guide to help plan for an interview as well as provide a record of a student's interview responses. The planning tool's elements help you decide how the interview can help you learn more about a student:

- Student understanding of your mathematical goal or goals for a lesson or several lessons
- Student conceptual understandings and/or procedural fluency
- Particular strategies used by the student during the interview (e.g., distributive property, doubling)
- The student's prerequisite knowledge or perceived challenges
- The student's mathematical dispositions

FIGURE 2.4 • **Planning for Interviews Tool**

Mathematics Goal(s): *Grade 3: Fluently multiply and divide within 100, using strategies such as the relationship between multiplication and division or properties of operations.*

ASSESSING	STUDENT RESPONSE	FEEDBACK TO STUDENT(S)	TEACHER COMMENTS/OBSERVATIONS
Conceptual Understanding	*Asked Matt to use graph paper to show related multiplication and division products and quotients using shaded rectangles*	*Reviewed and provided oral and written feedback to Matt's rectangular region drawings and how they modeled multiplication and division*	*The graph paper activity worked well. Will use this again. Matt did well here--took a bit more time than I thought, so we couldn't do that much, but I know he can do this independently--without me.*
Procedural Fluency	*Also asked Matt to orally state responses to triangular flash cards giving both multiplication and division responses (e.g., for 4, 5, 20; 4 × 5 = 20; 5 × 4 = 20; 20 ÷ 4 = 5; 20 ÷ 5 = 4)*	*Indicated which were correct or not orally, made a list of combinations needing further review*	*Matt was pretty quick and mostly accurate with the combinations reviewed. Will need more time for some combinations (6, 7, 42; 6, 8, 48; 8, 7, 56; 3, 8, 24)*
Strategies Used	*Matt didn't indicate the use of any strategies.*		*Need to ask Matt about his use of the commutative property, since he seemed not to understand how it could be used*
Student Prerequisites and Challenges	*No Challenges observed. This is early work on multiplication, so I didn't observe any lack of prerequisites.*		
Disposition	*Seemed focused for most, but not all, of our time together*	*Talked to Matt about staying focused on the work*	

General Comments: *Matt is doing well here. My next step is to provide him with an independent activity on the related facts he seemed to have difficulty with and those we didn't have time for. I will just observe how he does with this next step.*

Source: Adapted from Larson, M. R., Fennell, F., Adams, T. L., Dixon, J. K., Kobett, B. M., & Wray, J. A. (2012). *Common core mathematics in a PLC at work: Grades 3–5* (pp. 145, 146). Bloomington, IN. Adapted version published in C. Suurtamm (Ed.) & A. McDuffie (Series Ed.), *Annual perspectives in mathematics education: Assessment to enhance teaching and learning*. Reston, VA: National Council of Teachers of Mathematics.

online resources A blank template version of this figure is available for download at **https://qrs.ly/wsetnnz**

We suggest that you complete and maintain a brief record of the student's response and your feedback to the student either during or right after the interview. The planning tool also provides space to record additional teacher comments or observations as well as general comments related to the interview and student responses. Note that archived interview comments will be helpful to your ongoing planning, as a guide to intervention efforts, and as a source of day-to-day progress that may be useful within parent–teacher conferences.

1. **Think About:** I wonder about planning a lesson and thinking about interviewing my students before I actually teach the lesson. I have *never* done that before. How do I do this?

Classroom Response: *I keep a copy of the Planning for Interviews Tool (Figure 2.4) right next to me when I plan my math lesson. I use it to look at my lesson (for tomorrow) and then think about what I might ask a student or students about what they are doing. I pay a lot of attention to whether my "how or why" questions are about conceptual or procedural understandings. This is a helpful tool. I have made 4 × 6 file cards of the Planning for Interviews Tool that I carry with me when I interview students. I make notes about student responses and my feedback and review these notes each night before I plan.*

The Classroom Interview Record

An adaptation of the Planning for Interviews Tool is the Classroom Interview Record. Consider this tool as a way to quickly record interview responses for a group of students. When completed, as shown in Figure 2.5, the Classroom Interview Record provides an at-a-glance view of group responses and potential needs.

FIGURE 2.5 • **Classroom Interview Record**

STUDENT	MATHEMATICS CONTENT FOCUS (GRADE 6)	MATHEMATICAL PRACTICE(S)	LEARNING TASK	HOW DID YOU SOLVE THAT?	WHY DID YOU SOLVE THE PROBLEM THAT WAY?
Fran	*Proportions*	*Reasoning, model with mathematics, use of tools*	*Solved the following proportion problem.* *A cross-country runner ran 3 miles in 24 minutes. How long would it take the runner to run 5 miles?*	*I used a number line and just divided the distance by 3 to decide how long it took the runner to run 1 mile and then just figured it out.*	*Not sure, but we have been using number lines a lot, lately.*
Celeste	*Proportions*	*Reasoning, model with mathematics, use of tools*	*Solved* *3 miles in 24 minutes* *= 5 miles in x minutes*	*I just solved for the missing number in the equation.*	*I thought "3 miles in 45 minutes—that's 15 minutes for each mile." That helped me when I solved the equation.*
Add more rows as needed.					

online resources A blank version of this figure is available for download at **https://qrs.ly/wsetnnz.**

2. **Think About:** How can I get a quick sense of where my students are by how they respond to my interview questions?

Classroom Response: *I use the Classroom Interview Record (Figure 2.5) every day. A quick review of the student comments helps me when I think about groups for the next day's lesson. Maybe more important, at least for me, is that I have the whole-class list in front of me on the record, and it keeps me alert as to who I have or haven't interviewed and who I have interviewed multiple times. I have found this to be a great record-keeping tool that allows me to track my interviews at a glance.*

The Individual Student Interview Prompt

The Individual Student Interview Prompt (Figure 2.6) is a helpful resource that you can adapt to fit your needs and use to quickly record student responses to "how and why" questions you may ask during a brief interview designed to extend what you might see as you observe your students doing mathematics. You can fill in student interview responses journal-style and also use it for planning the next day's lesson or planning for intervention. Note that student interview responses are essentially student-to-teacher feedback related to your prompt—the interview questions.

FIGURE 2.6 • **Individual Student Interview Prompt**

INTERVIEW PROMPT*		
Name: Anna	**Date:** 3/4	**Math Topic:** Rounding Decimals to the Nearest Hundredth (e.g., 1.845)
QUESTION		**STUDENT RESPONSES**
1. How did you solve that?		I used base ten blocks to think through a couple of the problems, then I just remembered what we talked about the other day when rounding to the nearest tenth.
2. Why did you solve the problem that way?		Mostly I just remembered what we did the other day, but the base ten stuff helped me get started.
3. What else can you tell me about what you did?		I was a little unsure at first, but the blocks helped and then I remembered what we did earlier.

Note: Attach completed work sample(s).

A blank version of this figure is available for download at **https://qrs.ly/wsetnnz.**

3. **Think About:** As I began to think about using the interview on a regular basis, I wondered about both how to keep a record of student responses to my interview questions and how I might be able to access my questions easily.

Classroom Response: *I like to use the Individual Student Interview Prompt (Figure 2.6) for particular students I might interview. During math teaching time, I have the prompt questions on my digital tablet so as I observe I can be ready to do a short interview with a few students. Sometimes I actually pull together a small group and just ask them all to tell me how they solved a problem. I record their responses using my tablet's video camera, which gives me a "look" at what they are doing math-wise, but also their socioemotional connection to their classroom work, which I have begun to recognize as an important element of all my formative assessments.*

4. **Think About:** I find myself naturally and spontaneously conducting individual student or small-group interviews as I observe what's going on within a lesson. How can I connect my observations and student comments?

Classroom Response: *My sense is that an interview with a student or small group of students is an extension of what I have observed. The interview deepens my understanding of what I observed. Actually, lots of times I just need to ask my students what they are doing. It helps me understand their thinking and really identify strength spots that I can then use to launch other learning activities. I regularly record interview comments so that I can see growth between observations and interviews.*

Classroom Observation–Interview–Student Representations Tool

The Classroom Observation–Interview–Student Representations Tool (Figure 2.7), which is to an extent an adaptation of the work of Smith and Stein (2018), focuses on connections between observing, interviewing, and student use of representations. Student use and facility with varied representations is an important component in a student's understanding of mathematical concepts—at any grade/course level. This tool allows you to track anticipated and observed student use of strategies, methods, and/or representations, as well as the actual student use of particular representations, and provides you with an opportunity to suggest the order in which you may want students to share their strategies and how they used them to solve problems. When using this tool to interview, you should draft *assessing questions* (to get a sense of what students know and are able to do) and *advancing questions* (to help move a student's thinking forward) during the planning phase of your lesson. Aligning these questions with anticipated and observed student strategies can be useful in posing interview questions in response to what you observe, and for promoting rich class discussions. See the following Growing Corn task for an example of the Classroom Observations–Interviews–Student Representations Tool in use. Note the student group responses to the task and the use of assessing and advancing questions (the interview questions) for each group, which can then be used to determine the order of student presentations of each group's task solution strategies.

Growing Corn

A corn farmer measured the height of the corn when the first leaf appeared. It was 15 centimeters high. When the corn was ready for harvest 75 days later, it measured $2\frac{1}{4}$ *meters high.*

- *How much did the corn grow?*
- *If it grew about the same amount each day, about how much did it grow each day?*

FIGURE 2.7 • **Classroom Observations–Interviews–Student Representations Tool**

RESPONSE: STRATEGIES, METHODS, AND/OR REPRESENTATIONS	ASSESSING QUESTIONS	ADVANCING QUESTIONS	WHO	ORDER
210 cm in 75 days 210 cm ÷ 75 days = 2.8 cm About 3 cm	• Can you tell me what you did? • How did you arrive at 3 cm as your final answer? • Why do you think your answer is correct?	• Can you explain your work to another group?	Ishaan, Jenna, Mila	4th
Diagram with 75 circles (days) and 2 tallies in each circle (possibly 150 tallies in total) 210 − 150 = 60 60 ÷ 75 = 0.8 2 + 0.8 = 2.8	• How did you come up with that? • Why did you subtract 150? Divide 60 by 75? • What does 0.8 represent?	• If you solved a problem like this again, is there a more efficient way you could try? Explain.	Molly, Kai, Habib	3rd
2 meters = 200 cm $\frac{1}{4}$ of 100 cm = 25 cm 225 − 15 = 210 cm $\frac{210}{75} = 2.8$	• How did you come up with that? • What does 2.8 represent?	• What does the $\frac{1}{4}$ represent in the context of the problem?	Suhan, Parker, Tony	1st
225 − 75 = 150 cm (Group says they're stuck)	• Does 150 cm represent how much the corn grew or how much it grew each day? Explain.	• Have you used all the important information given? • Where do you think you are stuck?	Henry, Kaleb, Javier	2nd

Source: Adapted from Smith, M. S., & Stein, M. K. (2018). *5 practices for orchestrating productive mathematics discussions*. National Council of Teachers of Mathematics.

TECHNOLOGY TIPS AND TOOLS FOR RECORDING INTERVIEWS

There are some digital tools that provide teachers with the means to capture student interviews, each of which has specific advantages. As with any digital tool that is used to collect student data, be sure to investigate and follow school district data privacy policies and practices, and communicate the privacy plan and purpose of your recordings with parents/guardians and students.

- As with observations, video-recording apps available on digital handheld devices, tablets, or laptops (Chromebooks) can be effectively used to capture and archive interview discussions with and among students. One video discussion app, Flip (https://info.flip.com), provides a safe method for students and teachers to record and view video discussions/interviews. These digital interview files can be easily shared with the whole class and can serve as a centerpiece for classroom discussion and debate, as well as the focus for professional learning conversations among teachers and teacher leaders.
- Some of the best tools to use include the following interactive whiteboard apps: Jamboard (https://jamboard.google.com), Whiteboard.fi (https://whiteboard.fi), and Explain Everything™ (http://explaineverything.com). All these tools allow the user to capture what is being represented and discussed during an interview on a digital whiteboard screen.

Video—Conducting Interviews

Video 2.5

http://bit.ly/3VkI2J0

Michele, a third- and fourth-grade mathematics teacher, moves around her classroom observing and then interviewing pairs and small groups of students. She records interview responses using the Classroom Interview Record (Figure 2.5). Michele jots down quick notes of what her students say. The interviews provide an opportunity for her students to be heard—to communicate their learning.

Now that you are engaged in learning about the potential of observing and using interviews as classroom-based formative assessment techniques, **think about and discuss** how frequently you have been doing this. Have you recorded what you observed or interview comments of your students?

USING INTERVIEWS IN *YOUR* CLASSROOM

When planning for the use of the interview in your classroom, recall that the interview extends what you might have observed. This means that just as you anticipate what you might observe as you plan a lesson, so should you consider particular elements of a lesson (e.g., a particular problem's solution, the first-time use of the number line to compare fractions, use of graphing techniques, early experiences with solving multistep equations) that, when observed, could suggest the need for an interview.

As you think about the structure of most, but not all, interviews, it may be just asking a student to respond to a question or two about how they solved a problem and why they used a particular solution strategy to solve it. That is, the interview is based on something you have observed or anticipated observing. We think of the following as "starter" statements and questions you may use to essentially turn what you are observing or have observed into an interview.

- How did you do that?
- Tell me why you did that (e.g., solved the problem) that way.
- Do you understand what you did?
- How would you explain what you did to a friend?
- I noticed that you stopped working on this problem a while ago. What happened?
- I noticed that you changed your answer. How come? Explain your reasoning.
- Is there another way to solve this problem?

Of course, these starter statements and questions may be adapted to specific lesson tasks and the focus of what you have taught.

You can also consider providing your own separate task or activity solely for the interview with particular questions about this specific interview-created task. For example, the class could be working on tasks connecting estimation and reasonableness to operations with fractions, and your specific interview task might involve a problem like this:

Lucia walked at least $1\frac{1}{2}$ miles each day. Would she walk more than 10 miles in a week? Tell me how you know.

Interview questions related to this task could include the following:

- How could you (or did you) use a drawing or tool to represent this problem?
- How did you solve the problem?
- Do you think your answer is reasonable? How do you know?
- Did you think this problem was hard? Easy? Why?
- Follow-up question: Would Lucia walk 50 miles in a month? How do you know?

Note that the specifically designed interview task and related questions would need to be prepared and thought about either prior to the lesson or, and this is often the case, when you want to dig deeper as a response to what you expect to observe or have observed within your lesson.

An important thing to remember is that as you begin asking interview questions to assess or advance a student's thinking, be sure to allow the necessary time for the student(s) to process and respond *without* teacher intervention. While such intervention-like assistance is always well intended, it sometimes, if not often, results in the teacher actually doing the mathematics rather than the student. Consider Robin's confession:

> "I love using interviews, but have I learned a lesson along the way. That lesson is about patience! When I first started doing 'how and why did you do this' kind of interview questions, I wasn't patient enough to have my students process what I asked and then provide a response (oral or written) of their own. So, what did I do? Well, I ended up telling them! One day it just came to me. If the student waited long enough, they probably figured out that I would tell or show them what to do. No learning on their part. As I write this, I am smiling. Now, I pose my question (e.g., 'Tell me how you did that'), and I wait for their response before I say anything. I get so much more out of my students from the use of the interview. Experience is a good teacher!"
>
> **—ROBIN, FIFTH-GRADE TEACHER**

Most importantly, the interview as a classroom-based formative assessment technique extends what you observe, or anticipate observing, as you continually strive to determine and monitor the progress of your students.

Video—Reflection on Interviews

Video 2.6

http://bit.ly/3VvRfys

Beth Kobett summarizes how Michele, a third- and fourth-grade mathematics teacher, uses interviews in her classroom. As Michele reflects on the interview responses of her students, she notes that using interviews is an important way for her to truly monitor student understandings as well as consider instructional next steps.

Think about an upcoming lesson. Based on what you expect to observe, what kind of interview questions can you plan in advance to help uncover student thinking and make decisions about your next instructional steps?

Time to Try

As stated throughout this module, your use of an interview often extends what you have observed. The use of brief individual or small-group interviews essentially provides you with student-to-teacher feedback about the mathematics your students are doing—and learning. Student interview responses essentially define and document their understandings, levels of proficiency, and instructional next steps. This module helps you practice using one of the tools to plan for and implement interviews and includes suggestions for documenting student interview responses. So, let's try this out! As with Time to Try in the Observations module, consider a mathematics lesson you may plan for and implement sometime soon. Using the table that follows, provide brief responses to each of the statements related to planning for and using interviews. Discuss your "plan" with a colleague.

STATEMENTS	RESPONSES
Mathematics/Lesson Focus	
Anticipation: What might you observe, in the proposed lesson, that may prompt an interview?	
Provide several questions that you may ask in a brief individual student or small-group interview.	
What tool (see Figures 2.4–2.7) do you think you will use or adapt for use with your interview?	
What about the timing of the teacher-to-student feedback you will provide to your student or group of students? Will this occur immediately after the interview or later in the day? *How can you ensure that the feedback supports what students know as well as what they are still learning?*	

SUMMING UP

Interviews provide you with an opportunity to dig deeper with regard to what you might observe within a lesson. As you plan for their use, think hard about the purpose of your interview(s), including what you expect to learn and how interview responses may impact your planning for the next day and beyond. Interview responses provide potential identifiers for differentiation decisions, such as differentiating the pace of your lesson and differentiation plans for student groups and individual students. Also recognize that your understanding of the mathematics focus of the day's lesson and knowledge of your students should help to identify any lesson's "hot spots" that may prompt actual use of the interview for particular students. Finally, regular consideration and use of the interview should augment your observations and help in recognizing the need to not only consider differentiation but also recognize potential mathematical strengths and challenges, explore areas of advanced understandings, and notice and address concerns related to student mathematical dispositions.

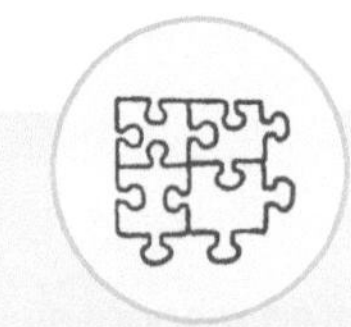

Your Turn

Rate, Read, Reflect! Consider and respond to the following questions with your grade-level or departmental teaching team or with teams across multiple levels (e.g., by grade level or course level).

1. As you regularly monitor student progress, how important is the use of interviews?

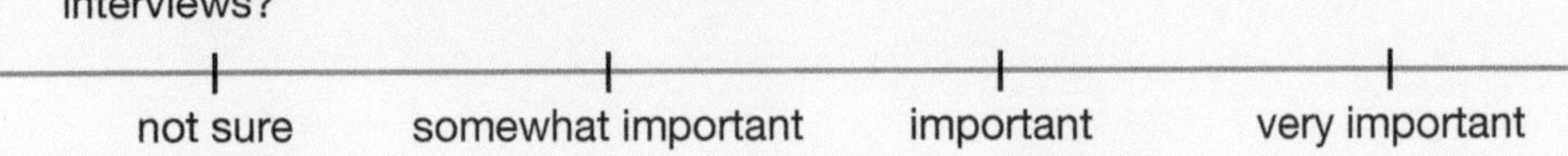

2. How often do you, or will you, use interviews to assess student progress in mathematics?

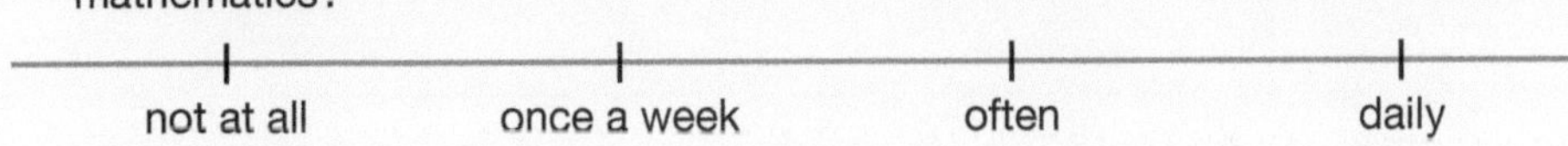

3. How frequently do you, or would you, respond to what you observe instructionally by conducting an interview with a student or small group of students?

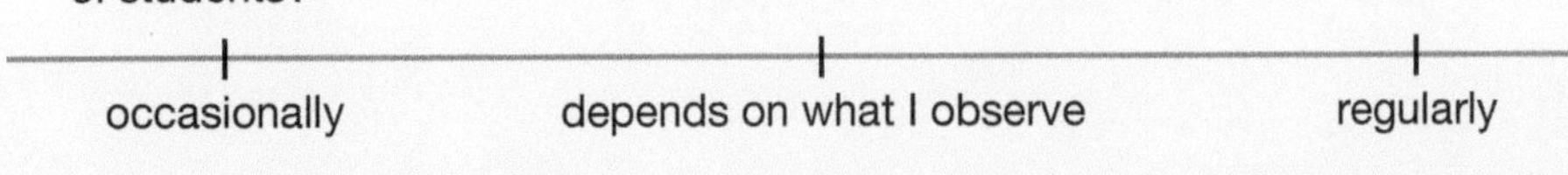

(Continued)

(*Continued*)

4. As you anticipate *when* you may interview one or more of your students, which of the interview tools presented in this module might you most likely use or adapt for use in your classroom?
5. How might you use the interview comments of your students in sharing their understandings and progress during a parent–teacher conference?
6. As you interview a student or group of students, how might you extend the interview to get a glimpse of student thinking beyond a particular content topic of interest or concern? Why would this be important?
7. What about feedback? As you listen to and note student responses while using the Individual Student Interview Prompt (Figure 2.6), consider how you might provide feedback to the student once you have had time to reflect on the student's interview comments. What's your plan relative to when and how you might provide feedback to the student?
8. Think about and create suggestions for preparing for, conducting, and analyzing comments and related responses (e.g., drawings) to an interview.
    - Preparing for the interview:
    - Conducting the interview:
    - Analyzing comments and related responses:

Now, discuss the following:

a. How could your suggestions be adapted for an interview with a small group of students?
b. How could your suggestions be adapted for an online interview?

NOTES

MODULE 3

SHOW ME

Grade 2: "Can you show me how you would order 76, 54, 47, and 89 using the number line?"

Grade 4: "How do you know $\frac{3}{4}$ is less than $\frac{7}{8}$? Show me."

Grade 8: "Show me your graph for that equation."

High School Algebra: "Show me and describe how you found the slope of that line."

FROM THE CLASSROOM

We were all talking about how we might use the Show Me formative assessment technique. Robin noted that she essentially uses it all the time, but never consistently records student responses.

Alex, who teaches at the middle school level, had never heard of the phrase "Show Me" before, but thought he should regularly incorporate the technique into his planning and teaching. Me, I just like the spontaneity of the Show Me technique. I do this all the time, but never thought about it as a way to formatively assess the progress of my students.

As I observe my students and even after conducting an interview, I often find myself just saying to a student, a small group of students, and even, on occasion, the whole class, "Show me how you would do X." I think such responses are particularly helpful as I try to assess my students' understandings of the use of representational tools, such as manipulatives and drawings.

We all agreed that using Show Me provides us with a student's performance, which we can then use as we consider next steps for needs related to our planning and teaching.

Purpose

You may have heard the phrase "show me—don't tell me." What does this mean, and how might it connect, in any way, to formative assessment? The Show Me formative assessment technique helps in providing depth to an observation or interview. It requires students to demonstrate their thinking—it's performance-based. You are now going to take a look at how to plan for the Show Me technique and consider suggestions that will help to make the implementation of Show Me effective and efficient in your classroom.

Module Goals

As you read and complete activities within this module, you will:

- ✓ Describe the connection between the Observations, Interviews, and Show Me techniques.
- ✓ Reflect on how you might use Show Me with individual students and small groups of students.
- ✓ Reflect on the value of Show Me as an assessment technique, particularly with student use of varied representations (e.g., visual models, drawings, manipulatives).
- ✓ Plan and teach a lesson incorporating the Show Me technique.
- ✓ Reflect on how you would provide feedback to students using the Show Me technique.

SHOW ME: BACKGROUND AND BASICS

Show Me is a performance-based response by a student or group of students that extends and often deepens what was observed and what might have been asked within an interview. Its use tends to be either serendipitous or planned. Teachers are often caught off guard by or wonder about a student's response within a lesson and can ask students to "show" what they did. Similarly, teachers can and should plan for particular elements within a lesson where a Show Me response may be warranted. Show Me activities often engage students in using and discussing representations (e.g., drawings, manipulative models, graphs, tables). Opportunities to communicate about such representations are particularly important for multilingual learners (Driscoll et al., 2016).

INSIGHT
The Show Me prompt requires a student or group of students to demonstrate their thinking and explain their response.

Like the Observations and Interviews techniques, when used regularly, Show Me has the potential to not only monitor but also improve mathematics teaching and learning. This is particularly true because use of the Show Me technique supports and encourages differentiation within activities that fully engage your students. Importantly, Show Me responses validate information gleaned from student observations and/or interviews and provide a heads-up for redirecting student responses within a lesson (Fennell et al., 2015).

Years ago, Sueltz and colleagues (1946) recognized that "observation, discussion, and interviews serve better than paper pencil tests in evaluating a pupil's ability to understand the principles used" (p. 145). We recognized that, not unlike teacher use of observations, many teachers have probably asked their students to show them what they were doing, perhaps without recognizing the assessment value of such performance-based opportunities.

Shavelson and colleagues (1992) noted that "a good assessment makes a good teaching activity, and a good teaching activity makes a good assessment" (p. 22). Having your students show what they are doing or what you request them to do is such an activity/assessment. Show Me is interactive in that its use can provide an indication as to the extent to which what has been taught has been learned.

When using the Show Me technique, students are engaged in presenting and, to an extent, validating the mathematics they are learning. *Principles to Actions* (National Council of Teachers of Mathematics [NCTM], 2014), in discussing beliefs about teaching and learning mathematics, notes that the role of the teacher rests with engaging students in mathematics activities that promote reasoning and problem solving and facilitate discourse. Student expectations, however, include being actively involved in making sense of the mathematics they are learning through the use of varied strategies, representations, connections to prior knowledge, or familiar contexts and experiences.

The NCTM's (1995) *Assessment Standards for School Mathematics* note two important purposes of assessment that relate directly to use of the Show Me technique. These include the use of assessment to both monitor progress and inform instructional decisions. The National Mathematics Advisory Panel (2008) recommended research regarding "think alouds" in mathematics, which is a strategy typically used in reading and somewhat similar to what we propose in this module as Show Me.

What About You? Show Me

- Think of a time when you were observing your students during a mathematics lesson, and you asked a student or a group of students to demonstrate or show their thinking; or perhaps you asked them what specific strategy or strategies they used as they were doing their mathematics. How was this helpful to you as a classroom-based assessment technique?
- React to the following statement from a teacher who regularly uses the Show Me technique:

  "I learned that when I ask one of my students to show me how they are using a particular computational procedure, solving a problem, using tools (e.g., manipulatives, drawings, the number line) to represent and solve a problem, and so on, I need to wait until they finish their entire 'Show Me' before I offer my comments. Early on, I would provide my feedback too quickly, which often shut the student down, discussion-wise, and just didn't provide me with enough about what they were doing—their strategies for solution—for me to know what they knew."

- The use of the Show Me technique, while often spontaneous or in-the-moment, can be anticipated. Think about and share particular mathematics lessons where the mathematics content and anticipated student responses would suggest use of the Show Me technique.
- How might observations of your students doing mathematics suggest the use of an interview or the Show Me technique? And do you think a student or small-group Show Me response might suggest that you interview a student or the student group—after the Show Me?
- How might you provide teacher-to-student and suggest student-to-student feedback when using the Show Me technique in your classroom?
- **Video:** Plan for and video-record Show Me responses of one of your students. Play the video back and provide two reactions to the student's responses during the Show Me activity. Share the responses with a grade-level team or mathematics department colleague.

Ponder This:

- How do you propose connecting the use of the Observations, Interviews, and Show Me techniques in your classroom?

PLANNING FOR SHOW ME

As noted at the beginning of this module, Show Me is a performance-based response. In many cases, it's essentially an on-the-spot extension of an observation or interview that deepens your understanding of what a student knows and can do, which then suggests next steps for your planning and teaching. Deciding to ask a student or group of students to show their work invites a demonstration of knowledge and understanding. Such responses help to identify and perhaps validate a level of progress in understanding or applying particular curricular standards within any of the mathematical content domains. Additionally, the very nature of a Show Me response will often address a number of Standards for Mathematical Practice (National Governors Association Center for Best Practices & Council of Chief State School Officers [NGA Center & CCSSO], 2010) or mathematical processes (NCTM, 2000). Show Me is essentially your planned or serendipitous extension of what you have observed or perhaps heard within an interview. Consider Erin's commentary on the use of Show Me in her classroom.

INSIGHT
Show Me is a stop-and drop, on-the-spot activity that should be relatively brief, less than five minutes.

"I was moving around the room and observing how my third-grade class was doing in creating drawings of a set of rectangles with the same perimeter, but different areas, and another set of rectangles with the same area, but different perimeters. I was watching Kira work and wasn't at all sure I could understand her thinking, so I asked her to show me rectangles with a perimeter of 12. She zipped right through this, showing 1 × 5, then 2 × 4, then 3 × 3 rectangles, noting that each had a perimeter of 12. And she talked me through each step of her work. However, when I asked her about the area of her rectangles, she wasn't at all sure about the area of any of her rectangles. I decided she needed a bit more of my time and that we would meet later, and I would have her make the same rectangles, all with a perimeter of 12, using grid paper or tiles. That way, I could have her literally see the square units and determine the areas of her rectangles. I was glad I conducted this interview and had Kira show me what she was doing and why. I also think Kira's response may be somewhat representative of others in my class. I need to think about how my lesson tomorrow can draw the connections between like perimeters and area and like areas and perimeter."

Erin's use of Show Me was essentially an extension of what she observed. It revealed that Kira seemed to understand the creation of rectangles with the same perimeter but was challenged by expressing the area of the rectangles. It also revealed the limitation of drawings and suggested the use of grid paper or tiles to represent the rectangles. Such models would provide the square units Kira may need to determine the area of the rectangles she created and deepen her understanding of the relationship between area and perimeter.

Let's consider this Show Me response. Think about how you might be able to use it if Kira was your student. Kira's responses truly impacted Erin's planning for her lesson tomorrow. Our thinking is that Show Me demonstrates a student's understandings, and also reveals potential learning opportunities that should inform instructional decisions. Importantly, such opportunities are most helpful in work with individual students or particular student groups. The questions in Figure 3.1 should be helpful to you as you both consider and complete preliminary planning for use of the Show Me technique, recognizing the brevity of such experiences. Later in the module, you'll find a tool (Figure 3.2) for planning and using Show Me as you teach.

Responses to the questions provided in Figure 3.1 should help you in considering the everyday use of Show Me as, essentially, a more in-depth extension of an observation or interview. Remember that, in most cases, Show Me will be a quick request for a student or perhaps small group of students to demonstrate what they are doing within the day's mathematics lesson. That said, you could also ask your entire class to complete a Show Me activity and do an observation of the responses to get a quick read of their level of strengths, possible mathematical challenges, or misconceptions related to the activity. You may sample students daily for Show Me opportunities or perhaps target students for Show Me based on particular needs, including as a quick assessment of the prerequisite conceptual and/or procedural knowledge necessary for the lesson of the day. Show Me is *your* opportunity to engage students in showing/demonstrating their conceptual understandings, procedural knowledge, use of particular representations or tools, and engagement of the Standards for Mathematical Practice (NGA Center & CCSSO, 2010) or the process standards (NCTM, 2000), particularly, but not exclusively, reasoning, representation, and modeling with mathematics.

FIGURE 3.1 • **Planning for Show Me**

1. Why would you use the Show Me technique?
• As you consider the use of Show Me, what particular aspects of your lesson would be most appropriate for having students show or demonstrate their understandings?
• What might you observe or expect to observe your students doing that might prompt you to use the Show Me technique?
• As you interview students in a particular lesson and consider the content and related instructional focus of the lesson, when might an interview transition to the more performance-based Show Me technique?
• Note: As you consider the importance of a regularly updated profile of the progress of your students in mathematics, the potential of Show Me to address individual needs is particularly valuable, given that you can suggest varied responses to the same Show Me prompt (e.g., "Use a drawing to show . . ."; "Use the number line to show . . .").
2. Where might Show Me opportunities occur within your lesson?
• How might you use Show Me as students are working independently within the day's lesson?
• How might you use small-group or whole-class opportunities for students to show you how and why they completed a particular activity before releasing them to independent activities?
• Can you think of particular students or groups of students who might benefit from regular opportunities to show you how and why they completed particular elements of a lesson's activities?

3. How will you organize your classroom to implement Show Me each day?

- As you consider your implementation of the Show Me technique, how might you design a particular location in the classroom for Show Me or take advantage of the at-their-seat, in-the-moment spontaneity for implementation of Show Me?
- How might you provide easy access to representation tools as needed for a Show Me response (e.g., drawings, technological tools, manipulative materials)?
- As with the Interviews technique, for particular areas of interest/concern, how might you record Show Me responses using a video or audio recording, a digital camera, or perhaps an interactive whiteboard app (e.g., Explain Everything™, Educreations, or ShowMe interactive whiteboards)?

4. As you implement Show Me, what initial questions/statements might you use as you seek a deeper understanding of student thinking? Consider the following as possible examples.

- Can you show me how you did/are doing that example and explain your reasoning?
- Show me how to solve that problem using a number line (or any other related representation tool).
- Show and share your solution strategy with me.
- Show and share your solution strategy with your team member, making sure to describe how and why you did what you did.

INSIGHT

As you consider teacher-to-student feedback, what might you ask or do as you consider a student's Show Me response? How might you recognize and acknowledge the strengths of the work the student showed you?

5. What might you anticipate from students? (Consider understandings, possible mathematical challenges or strengths, and extensions of the mathematics presented, as well as evidence of student disposition.)

- As you plan the lesson, before engaging students in the Show Me activity, reflect on your class and its progress, and anticipate the types of responses you may receive. Would the Show Me prompts used reflect conceptual or procedural understandings? Would your students be able to restate, orally or in writing, their reasoning?
- How might you consider selecting a sample of students in your class for Show Me each day, using the responses to monitor and advise your planning as well as identify individual student needs? The sample should be representative of the achievement range within your class, including the various ways your students demonstrate understanding from early, naïve conceptions to fully developed conceptual and procedural understanding, as well as students whose achievement seems to vary from week to week or even day to day.

6. How might you follow up after a Show Me request?

- If additional Show Me follow-up is considered, would it consist of additional examples?
- How might Show Me responses influence your planning for the next day's lesson?

Time Out

Let's Reflect:

- As you plan and anticipate teaching a lesson, what particular challenges content-wise or instructionally may suggest your use of the Show Me technique?

____

____

____

____

- Student use of particular representations (drawings, expressions, equations, manipulatives, number lines, graphs, etc.) is a frequent Show Me opportunity. How might you organize a Show Me request that engages a representation frequently used at your grade level (elementary/middle) or within one or more of the courses you teach (middle or high school)?

____

____

____

____

- Consider the following comment from a teacher who regularly uses the Show Me technique: *"While I regularly use Show Me as I observe my students, I have also found that just stopping a lesson and asking the class to show me (on paper or a tablet device) their solution strategy, and then moving quickly around the room to check responses, is a good way for me to literally 'see' what my class is doing, and then move forward with the lesson, stop and review, identify students for interviews, and so on."* Would you consider such full-class use of Show Me?

____

____

____

____

(Continued)

(Continued)

Ponder This:

- Based on your reading and engagement with the Show Me technique in this module, how will *you* use Show Me in your classroom? Provide one or two examples of what you anticipate your students showing.

____

____

____

____

- Recognizing that an individual student, small-group, or whole-class Show Me response to a prompt you provide (e.g., "Show me how you solved that problem"; "Use a drawing to show me the difference") is essentially student-to-teacher feedback, which will elicit your teacher-to-student feedback, what will be important for you to record (notes, picture, video) as a reminder of the student's Show Me response?

____

____

____

____

Video—Using Show Me in Middle School

Video 3.1

http://bit.ly/3EUupcH

Skip Fennell describes the importance of and use of the Show Me technique. A middle-grade lesson taught by Joanne engaged a whole-class Show Me opportunity as students, working in groups, solved a problem related to estimating the number of old and young trees on a large tree farm. Joanne moved around her classroom asking her students to show her the strategies they used to solve the problem, encouraging student-to-student feedback within the groups and also providing her own feedback to the students. Joanne discussed how using Show Me allowed her to document the student responses and very quickly gauge their progress in the activity.

Think about one of your own classes. What did you see in this snapshot of Joanne's whole-class use of the Show Me technique that you might also implement as you use the Show Me technique?

TOOLS FOR USING SHOW ME IN THE CLASSROOM

As you plan for the daily use of the Formative 5, it will be important to consider how the trio of related assessment techniques—Observations, Interviews, and Show Me—can be used to not only monitor your instruction but also provide you with feedback as to the relative success of your planning as well as student progress. As you think about what you will observe and when you might interview your students, you will also consider particular opportunities for using the Show Me technique.

Consider the following classroom examples as merely a sampling, with actual classroom vignettes provided, of why you might ask students to show what they are doing or have done within a particular mathematics lesson.

1. **Think About:** As you observe a written or hear an oral student response, use Show Me to understand what was done and why.

   Classroom Response: *Javier, a fifth grader, said the following to his mathematics partner: "If I multiply a whole number, say 2, by $\frac{1}{2}$, the product is 1—it's smaller! But if I divide a whole number, let's use 2 again, by $\frac{1}{2}$, the quotient is 4—it's larger." What would you ask Javier to show you to indicate his understanding of his statement?*

2. **Think About:** Use a sampling of Show Me requests, individual student and/or small group, to determine how students complete an activity integral to the day's lesson.

   Classroom Response: *In Amanda's class, her third graders worked in small groups. In today's lesson, she began by reviewing the students' work with multiplication and factors of numbers using area models. She had student pairs use square tiles or drawings to help determine a response to the following task:*

   > Our physical education teacher wants to retile an area of our gym using 1-foot tiles. She has 36 tiles, and the area to be tiled is rectangular. Using your square tiles or drawings, show me all possible measurements of the gym's tiled area. Then decide which rectangular area you would recommend to the physical education teacher for tiling, and justify your decision.

   As she observed her students completing this activity, Amanda noted that Carlos and his partner, Diago, were discussing this problem, orally stating that, "hey, since the area is rectangular, it has four sides, so we can just think 4 times some number is 36 and we've got it." Sam and Paige just used their tiles and made different rectangular arrangements, with drawings of each, showing the following lengths and widths: 1 × 36, 2 × 18, 3 × 12, 4 × 9, and 6 × 6. Amanda asked Carlos and Diago and Sam and Paige to share their solutions to the task and provide feedback to each other's solution, using the shared responses to help launch the day's lesson, which focused on continued work with factors.

3. **Think About:** Use Show Me to monitor the progress of particular students who may lack mathematics prerequisite knowledge or have missed instructional time on a topic.

   Classroom Response: *Jason strategically used Show Me to monitor the progress of two of his fourth-grade students. Tony had missed close to a week of school due to illness. Jason was using Show Me each day with Tony to help determine his progress from day to day. He found it particularly helpful to ask him to show how to divide, beginning with problems like 63 ÷ 5 and then moving to dividing a 3-digit number (e.g., 124 ÷ 7). Tony used base ten materials to represent the division process. Jason then had him create and solve two word problems (a 2-digit number divided by a 1-digit number and a 3-digit number divided by a 1-digit number) to provide a context for his division examples, making sure that Tony discussed his reasoning when solving the problem. Jason felt that Tony was growing in his confidence and understanding of how to use the base ten manipulative tools to represent division problems and their solution as well as connecting division to real-life contexts that made sense to him.*

   Jason wanted to learn more about Nancy's understanding of division and asked her to show how she would use the base ten materials to solve both 87 ÷ 4 and 167 ÷ 9. This was helpful since he noted that Nancy seemed quite facile in her use of the materials to solve the problems but was quite challenged when asked to create word problems to show a context for each of the problems he had just solved—computationally. Jason decided that it made more sense to regularly use word problem contexts to introduce all additional lessons and practice opportunities involving division, making sure to engage his students in both representing the problem's solution and discussing their reasoning.

4. **Think About:** Use Show Me to challenge student thinking. As students successfully demonstrate progress/understanding of the day's lesson focus, be prepared to use Show Me to challenge those students by extending the mathematical focus of the lesson.

   Classroom Response: *Nia observed two of her sixth graders, Deion and Tavon, as they were organizing and thinking about a way to represent the data collected from a class survey. She noted that Deion and Tavon had organized the survey data as they began creating a line plot to display the survey data. She asked them to show her how their data (the line plot) would look if they found 10 more survey responses for the lowest line plot data point and to discuss what they would now say about the survey results. Nia felt that it was important to regularly plan for such lesson or task extensions as possible Show Me activities for her students.*

5. **Think About:** Identify understandings and use a sampling of Show Me responses to advise planning for the next day's lesson and the identification of possible small-group needs.

Classroom Response: *Susan sampled at least five students every day with a Show Me prompt. She then carefully considered the range of these responses to help guide her planning for the next day with possible implications for content, pace of the next day's lesson, and use of a Show Me activity to define small-group activities within her lesson.*

6. **Think About:** Be prepared with a Show Me request that would extend a student's knowledge beyond the focus of the day's lesson.

Classroom Response: *Anthony liked to use Show Me with a selected small group of students to get a glimpse of what he might do in tomorrow's lesson. Today, his algebra students were determining the relationship of the data represented within a table and then using Show Me to describe in words and by writing an equation the pattern represented by the table. Tomorrow, he hoped to have them represent a linear function using a table and making a graph to show the pattern represented by the table. He gathered Manuel, Leslie, and Ian and asked them to look for a pattern that they could describe in a table of miles traveled and gallons of gas in the tank of a school bus. Using Show Me, he then asked them to represent the pattern represented in the table using words, an equation, and making a graph. He hoped they would see the connection between the tabled data, equation, and graph, which would provide a good launch to discussing functions and extending work on this topic. Anthony felt that these preplanned Show Me activities and the student responses were most helpful to him in jump-starting his planning for the next day.*

FIGURE 3.2 • Small Group: Show Me Record

<table>
<tr><td colspan="2">SHOW ME: Mathematics Content: Grade 2: Number and Operations—Base Ten
• Task: Use the base ten blocks to create a value that meets the following conditions:
• Closer to 400 than 300
• Even number of tens
• Odd number of ones
• The digit in the ones place is greater than the digit in the tens place</td></tr>
<tr><td>Lesson Focus/Standard: Understand that the three digits of a three-digit number represent amounts of hundreds, tens, and ones (e.g., 706 equals 7 hundreds, 0 tens, and 6 ones).</td><td>Anticipated Student Show Me Responses:
Some students will create a number value and then check it against the conditions.
Other students will create the value as they read through the conditions.</td></tr>
</table>

(Continued)

(*Continued*)

Student: Nick	**Student:** Ella
 "I show that my tens are even because I grouped my tens by twenties in different colors! My total is 385!"	 "I put 4 hundreds blocks down because I knew I wanted to get as close to 400 as possible. Then, I covered up the 3 hundreds blocks to show 300 and then put 10, 20, 30, 40, 50, 60, 70, and 80 down, and then 9 ones, to make 38."
Student: Annabelle	**Student:** Derrick
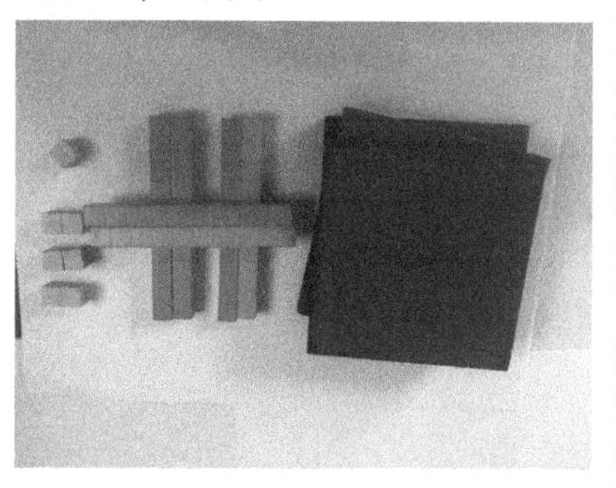 "I made 367. I have 3 hundreds and then 6 tens. Do you see how I grouped them so you can see they are an even number? Each ten has a partner ten."	 "I made 389. That's 3 hundreds, 8 tens, and 9 ones."

A blank version of this figure is available for download at **https://qrs.ly/wsetnnz.**

The tools and student responses to the following Show Me prompts will be helpful to you as you consider the everyday use of Show Me in your classroom. The Small Group: Show Me Record (Figure 3.2) provides a recording of Show Me responses from a sampling of students based on the needs/uses addressed earlier. This tool helps in organizing and providing an actual record of student responses, which you can then consider as you plan for the next day's lesson. Given the performance nature of Show Me, we have found that pictures of student responses along with a brief, accompanying notation are a quick way to document student Show Me responses. Access the tool in Figure 3.2 for your use at **https://qrs.ly/wsetnnz**. It should also be noted that this tool could easily be adapted as an individual student Show Me.

The classroom examples from Mr. Cardoza and Mrs. Hu that follow both involved use of the Explain Everything™ interactive whiteboard app (https://explaineverything.com) as a digital Show Me tool. The responses to particular tasks and the teacher suggestions as to how the Show Me responses were both interpreted and used for planning are included.

* * *

Students in Mr. Cardoza's fourth-grade class provided responses to the following problem:

Antonia's family decided to drive to Ski Round Top, which was about 680 miles from their home. Jon had a much shorter drive, which was about $\frac{1}{2}$ the distance that Antonia's family drove. How much farther did Antonia's family have to drive?

As Mr. Cardoza observed students, he selected Annika and Chayse to show their representations using the Explain Everything™ whiteboard app. Annika's response, involving a mental mathematics approach, is provided in Figure 3.3. She decomposed 680 into 600 and 80, and then broke 600 into two groups of 300, and 80 into two groups of 40, arriving at 340 miles. Mr. Cardoza made note of her representation and Annika explained, "Half of the distance that Antonia's family drove is Jon's amount. And Antonia's family drove 680 miles, and half of that is 340 miles."

FIGURE 3.3 • **Annika's Work**

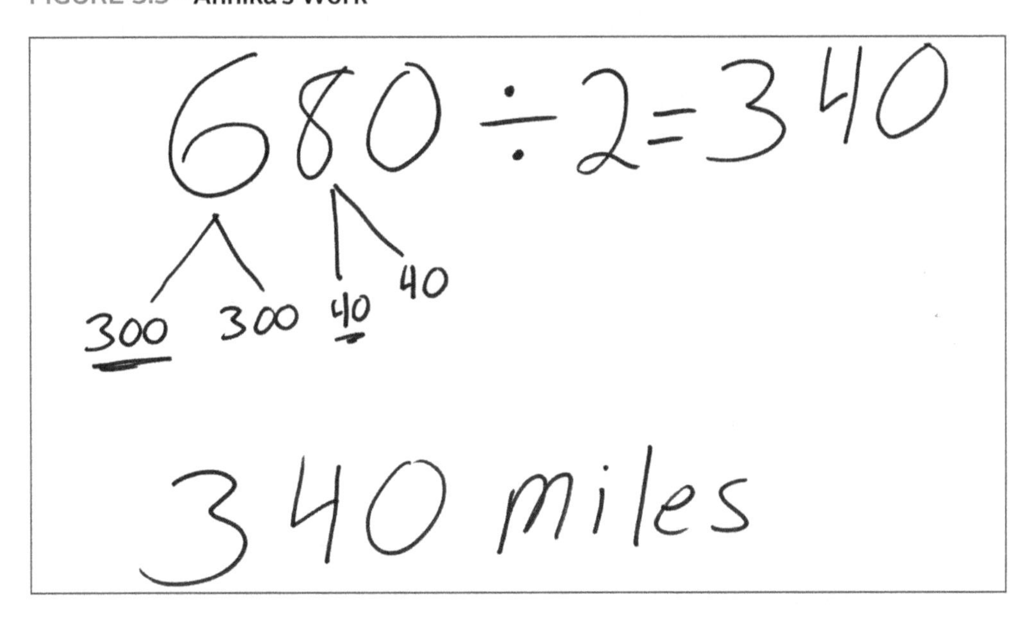

Video 3.2
Annika's Work

Mr. Cardoza then observed Chayse's work. Chayse had also translated the problem into dividing 680 by 2 and used the division algorithm (see Figure 3.4). When Chayse discussed his Show Me response, it revealed his ability to decontextualize, using numbers and symbols to explain the steps in the procedure for dividing 680 by 2.

FIGURE 3.4 • **Chayse's Initial Work**

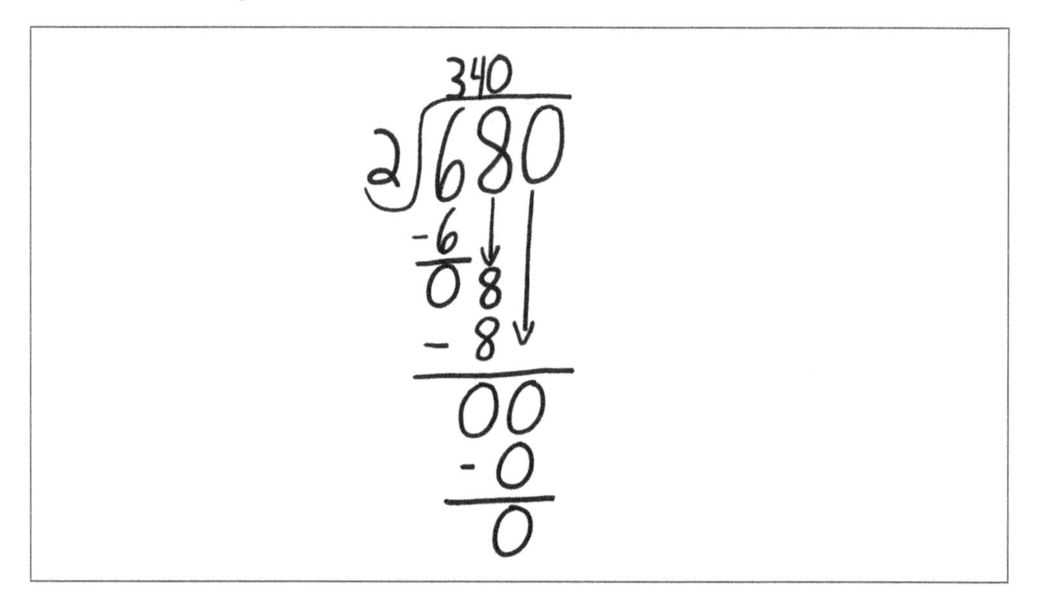

Mr. Cardoza wondered if there was any evidence that Chayse could contextualize—or make sense of—the quantities and relate them to the problem involving distance. Mr. Cardoza asked Chayse if he could show his reasoning using a picture or diagram. Chayse's diagram is provided in Figure 3.5. He explained, "Okay, you have to have this house on the left . . . and [Antonia's] going to Ski Round Top—say that's all the way over here [draws a horizontal line]. Jon's house would be half of that, which is right here [draws a house at the halfway point on the line]. This would be [and labels] Ski Round Top." This model could work, but does Chayse realize that Jon's house might not be directly between Antonia's and Ski Round Top?

FIGURE 3.5 • **Chayse's Show Me**

Video 3.3
Chayse's Initial Work

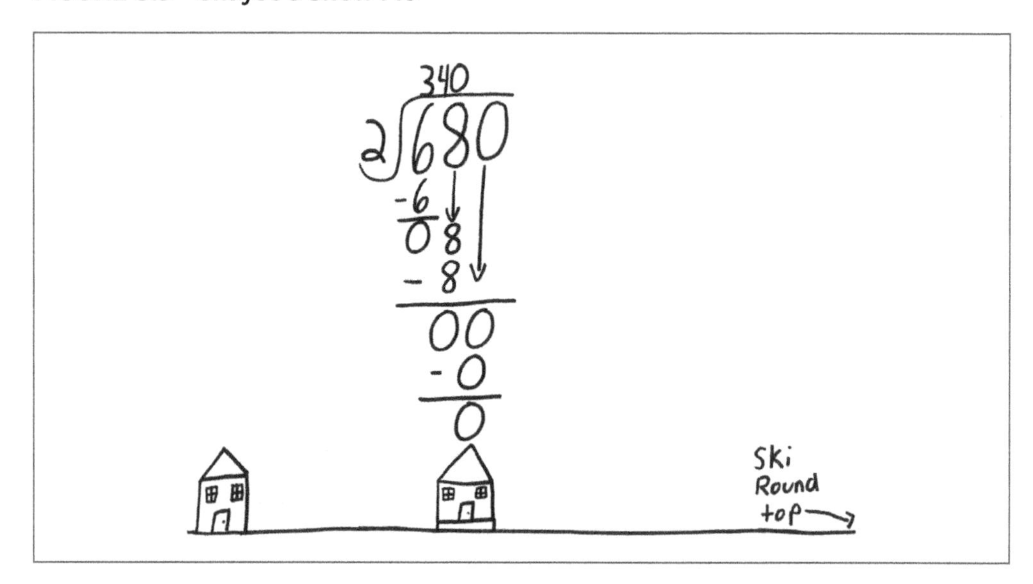

Mr. Cardoza observed other students as they worked and collected each student representation for that problem. He realized that most students solved using methods and representations similar to Annika's and Chayse's (using mental mathematics or the division algorithm). Mr. Cardoza reflected on the effectiveness of using a Show Me prompt with his students as a technique to help get a real sense of whether or not they can contextualize, understanding that 340 miles represents both the halfway point (got it) as well as how much farther Antonia had to drive than Jon (maybe, but not completely sure). He planned to follow up with a full-class discussion involving an examination of student representations such as Annika's and Chayse's, followed by questions designed to get a sense of whether or not students understood "340 miles" as the value of how much farther away Antonia's house was from Ski Round Top than Jon's.

* * *

Mrs. Hu's seventh graders worked on the following task:

Cam took 15 shots and made 9 of them, scoring 18 points. He had the same shooting percentage in his next game; how many shots could Cam have made in that game?

Alanna, Jordan, and Kyra also shared their reasoning using the Explain Everything™ app. Figure 3.6 shows how Alanna divided $\frac{9}{15}$, the greatest common factor of 3, and used $\frac{3}{5}$ to generate an equivalent fraction $\left(\frac{6}{10}\right)$ to find Cam's shooting percentage (60%).

FIGURE 3.6 • **Alanna's Work**

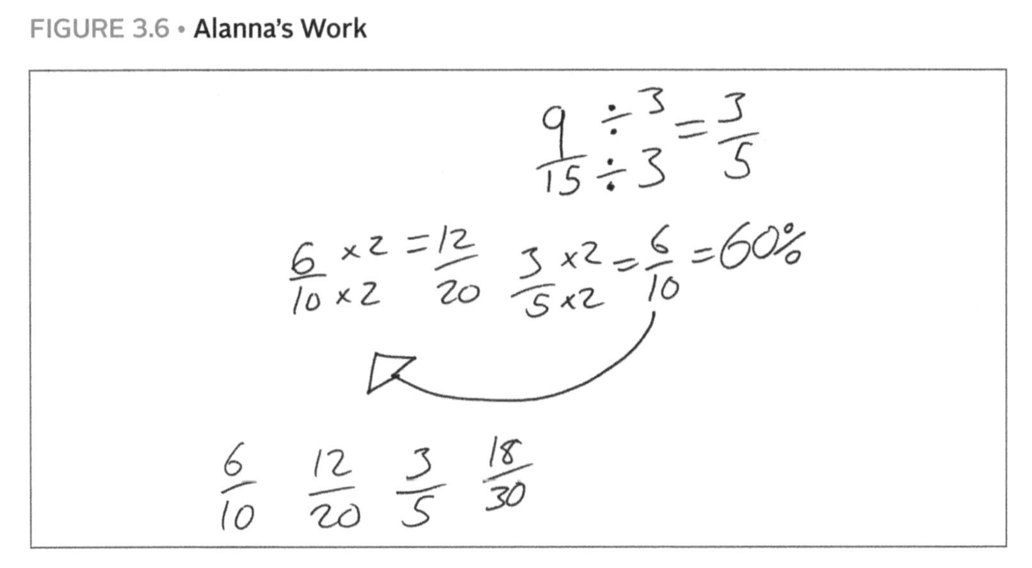

Video 3.4
Alanna's Work

She used common factors to find other equivalent fractions and stated, "He could have gone 6 for 10, 12 for 20, 3 for 5, or 18 for 30. Basically, just anything that equaled 60 percent." It's worth noting that having the animated, audio version of Alanna's solution provided Mrs. Hu with a well-articulated learning artifact that helped her get a true sense of Alanna's procedural and conceptual understanding—one that could not have been gathered as easily without asking Alanna to record her Show Me using the digital app.

Mrs. Hu observed Jordan's initial solution, 60% (Figure 3.7). When asked to explain, he stated, "Nine divided by 15, that equals 60%." She agreed and asked

Jordan if he could show how many shots Cam could have made using other fractions equivalent to 60%. Jordan found that $\frac{3}{5}$ and $\frac{6}{10}$ were both equivalent to 60%, and although concisely stated, he provided Mrs. Hu with evidence indicating his understanding.

FIGURE 3.7 • **Jordan's Work**

9 ÷ 15

$15\overline{)9.0}$ = .60   60%
−9.0
0

$\frac{9}{15} = \frac{3}{5}$

$\frac{6}{10} = 60\%$

3 ÷ 5 60%

Video 3.5
Jordan's Work

Kyra's approach (see Figure 3.8) focused on trying to find a proportion using the cross-multiplication algorithm. She determined that Cam's shooting percentage was 60% and concluded that he "would have made 6 shots if he shot 10 times."

FIGURE 3.8 • **Kyra's Work**

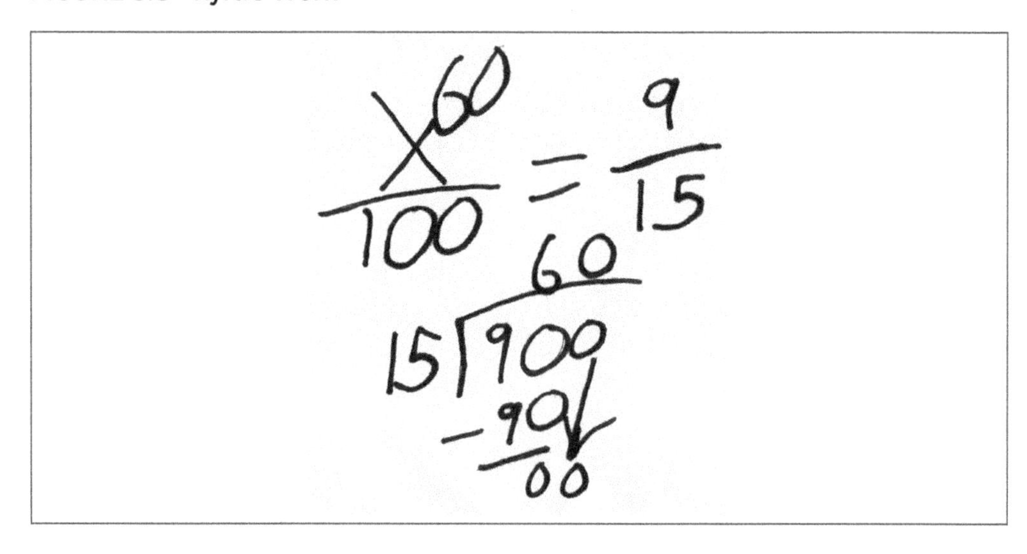

Video 3.6
Kyra's Work

Mrs. Hu decided that a follow-up interview would be her next step with Kyra as she wanted to dig a little deeper into the meaning behind the procedure Kyra used to find Cam's shooting percentage. How were those fractions related? Why does the cross-multiplication algorithm work? Can Kyra come up with other possible solutions (the actual shots made to be equivalent to 60%)?

* * *

The students in Mrs. Newsome's Algebra 2 class had been reviewing arithmetic and geometric sequences prior to working on the following task (adapted from Illustrative Mathematics, Algebra 2) involving exponential growth:

In a photograph of Elena, the distance from her chin to the top of her head is 150 mm. For a passport photo, this measurement needs to be between 25 mm and 35 mm. How many times would the image need to be scaled down by 80% for the distance between her chin and the top of her head to be less than 35 mm?

Pairs of students worked together on the task, while Mrs. Newsome observed them engaging in problem solving and discussions. She compared their work with what she had anticipated they might do in response to the task while planning for the lesson. Figure 3.9 shows Molly and Lorenzo's first attempt.

FIGURE 3.9 • **Molly and Lorenzo's Work: Part 1**

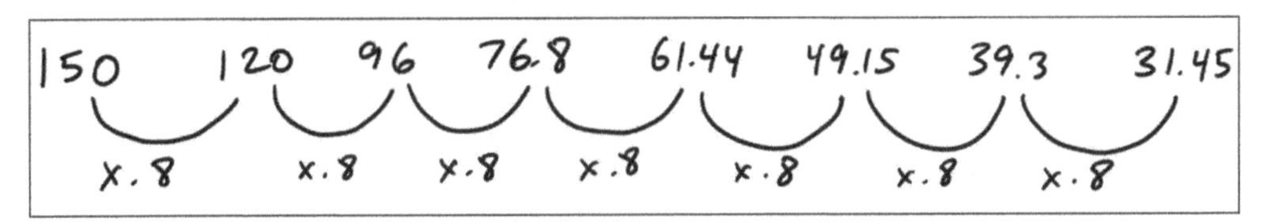

Mrs. Newsome studied their work. "Can you tell me about your mathematical ideas?" she asked.

Molly replied, "We started with the original height of Elena's picture and multiplied by 0.8, or 80 percent, until we got a value less than 35 millimeters."

In an effort to gain more insight into their understanding of the solution, Mrs. Newsome posed the following Show Me prompt: "Can you show me your ideas using a picture?" Molly and Lorenzo got back to work. When Mrs. Newsome revisited the pair, she observed their work, shown in Figure 3.10.

FIGURE 3.10 • **Molly and Lorenzo's Work: Part 2**

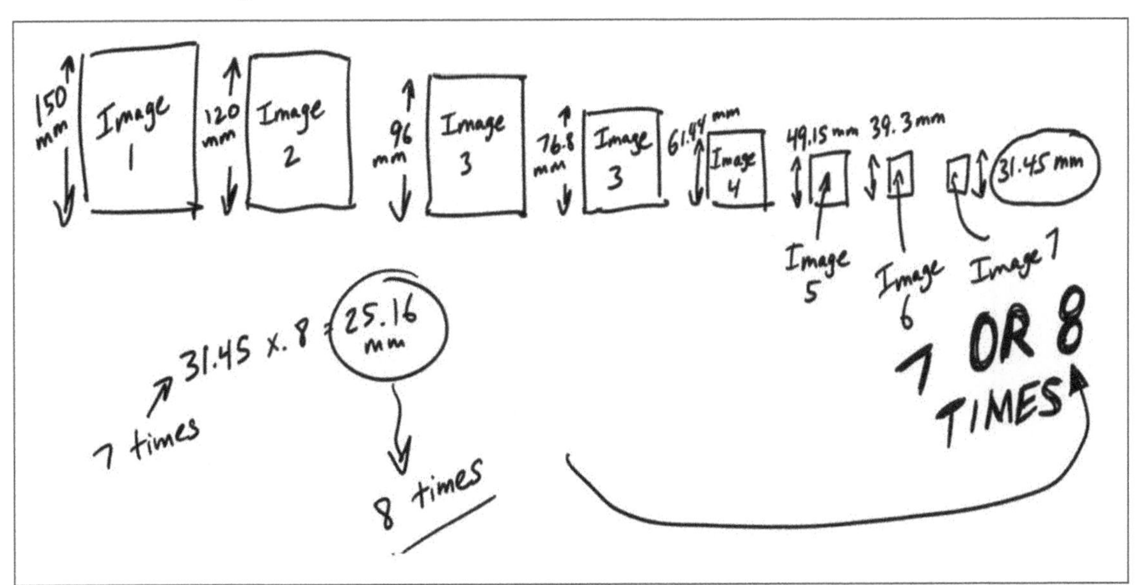

Lorenzo explained, "We decided to draw images of the photo, beginning with a height of 150 that showed how Elena shrunk it by 0.8, or 80 percent, each time, like in each of our trials."

Molly added, "And we also figured out that the answer is seven *or* eight trials, because when we shrunk the photo by 80 percent an eighth time, it was still just above 25 millimeters."

Angela and Rob had created a similar representation (see Figure 3.11) as Molly and Lorenzo's initial attempt. Mrs. Newsome listened to their conversation, which focused on a review of geometric sequences. Rob said, "I know that we can write this using exponents." "Right," Angela commented. Mrs. Newsome listened, paused, and then asked, "So is it possible for you to generalize the work from your table, by showing me your thinking in the form of an expression or equation?"

Angela recalled that a geometric sequence was at play in this scenario and said, "If you multiply the initial number, like 150, by 0.8 to the power of the number of trials in the sequence, you should get the new size of the photo."

Mrs. Newsome asked, "Can you show me by using function notation?" Rob and Angela discussed their ideas, and when Mrs. Newsome returned, they had generated an additional solution as shown in Figure 3.12. Mrs. Newsome's students demonstrate how the Show Me technique truly has the potential to reveal both the importance of representation and levels of student thinking.

FIGURE 3.11 • **Angela and Rob's Work: Part 1**

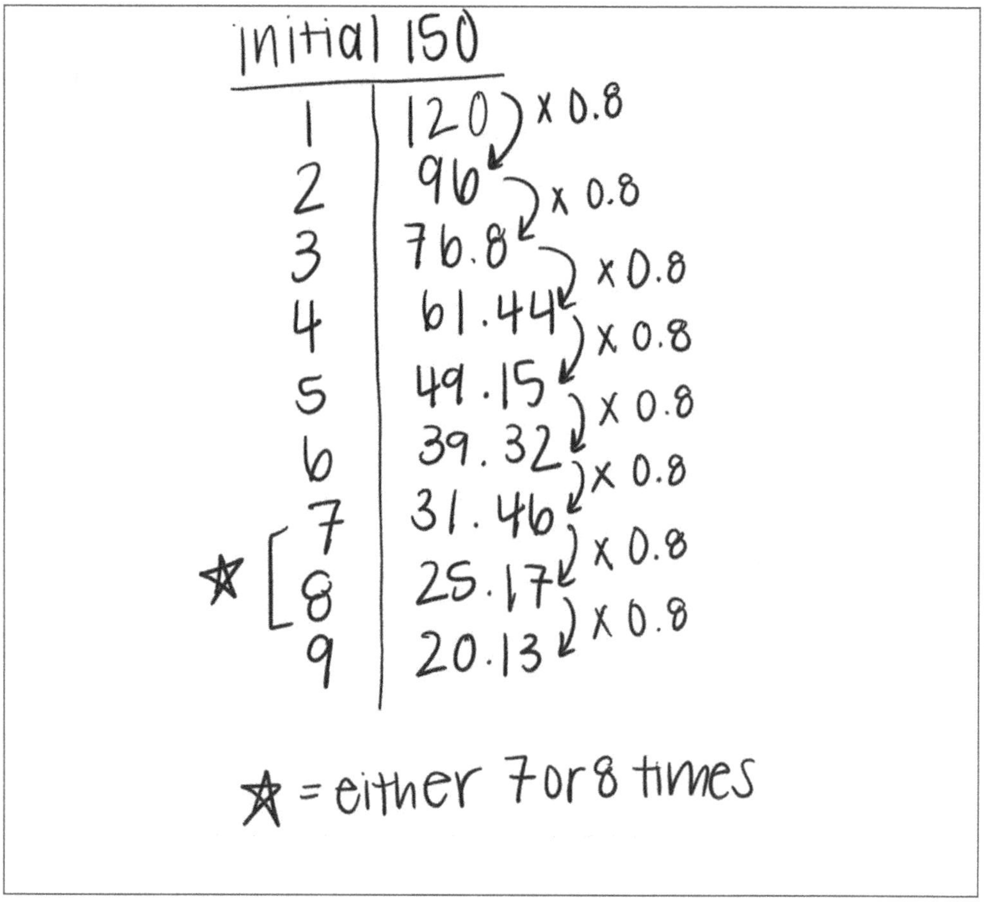

initial 150

1	120	x 0.8
2	96	x 0.8
3	76.8	x 0.8
4	61.44	x 0.8
5	49.15	x 0.8
6	39.32	x 0.8
7	31.46	x 0.8
★ 8	25.17	x 0.8
9	20.13	

★ = either 7 or 8 times

FIGURE 3.12 • **Angela and Rob's Work: Part 2**

$f(n) = 150 \times 0.8^n$

$n = 7$

$f(n) = 31.46\text{mm}$

$n = 8$

$f(n) = 25.17\text{mm}$

The Show Me responses by Mr. Cardoza's, Mrs. Hu's, and Mrs. Newsome's students demonstrate the valuable contributions such opportunities provide for impacting planning and monitoring and guiding mathematics teaching and learning.

Focusing on Feedback

Like a student's response to an interview, a student's response to a Show Me prompt is student-to-teacher feedback. It's a response to your prompt. Your "on your feet" analysis of a student's Show Me response will certainly inform your teacher-to-student feedback. But let's think about the reality of such Show Me responses, the importance of your analysis of their response, and the feedback *you* provide to your students. Consider each of the following statements and suggest your feedback. Make sure that the feedback you provide will help your students learn more about their thinking.

a. A response that **is provided so quickly that you actually wonder how the student "did that."** What's your feedback?

b. A response that **you just don't understand.** What's your feedback?

c. A response that **uses multiple representations (e.g., drawing, table), but was not the response representation-wise expected or requested.** What's your feedback?

d. A response that **doesn't really "show" but is an oral statement that seems accurate.** What's your feedback?

(Continued)

(Continued)

e. A student response that **is partially correct (and partially incorrect).** What's your feedback?

f. A response that **reveals a student strength, beyond what you had anticipated, when planning for use of the Show Me technique.** What's your feedback?

g. A response to a class-based Show Me prompt **where seemingly half of the class provides correct responses, and the other half does not.** What's your feedback?

TECHNOLOGY TIPS AND TOOLS FOR RECORDING SHOW ME RESPONSES

There are digital tools that provide teachers with the means to capture student Show Me responses, each of which has specific advantages. As with any digital tool that is used to collect student responses, be sure to investigate and follow school district data privacy policies and practices, and communicate the privacy plan and purpose of your recordings with parents/guardians and students.

- Like other technological tools and techniques, the previously mentioned interactive whiteboard apps and digital video-recording devices and apps can be used to capture the observation of students engaged in a Show Me performance. These recordings allow teachers and students to document and review student problem solving as they are engaged in the mathematics they are learning and help inform planning for next steps.
- A more comprehensive tool designed to improve critical thinking and communication skills, while also serving to capture and share student representations, models, and solutions, is CueThink (http://cuethink.com). Teachers select or create their own problems, posing them to students inside the application. Students use the oft-mentioned four-step problem-solving model—Understand, Plan, Solve, Review (Polya, 1945)—and have the opportunity to consider what they notice and wonder about the problem, generate an estimate, and choose an initial strategy or plan for solving the problem. Students create "thinklets" or video representations containing drawings, images, and audio using the intuitive drawing and writing tools provided within the application. Students encounter quality checks and reminders to consider as they prepare to submit their solutions. The true power of CueThink comes into play when classmates are able to view, comment on, critique, and add annotations to one another's thinklets. These Show Me student artifacts can serve as portfolios of student reasoning and problem solving.

There are a few important considerations related to using digital tools to capture a Show Me performance. Careful thought should be given to each of the following.

1. Intended purpose: Is the goal to capture and archive or distribute digitally to classmates and/or the teacher?

2. Function and ease of use of the digital tool or application: Does the tool provide an opportunity to capture information about student thinking in a way that significantly enhances what could be done without it? Can students access the tool efficiently and generate flexible representations that provide the teacher with feedback related to student understanding?
3. Method(s) for gathering Show Me responses and providing students with feedback: Students who have one-to-one access to devices (e.g., laptops, tablets) have a considerable advantage in terms of being able to generate and share their ideas electronically. However, a teacher with only one (or a few) digital device(s) (e.g., iPads, laptops) can, during an observation, make a Show Me request to an individual student or small group of students. The digital file can easily be saved, shared, analyzed, and discussed as a classroom learning artifact, serving as a resource to help improve learning for all students.

USING SHOW ME IN *YOUR* CLASSROOM

When planning for the use of the Show Me technique, remember that this is a "stop-and-drop" activity where you will ask a student, a pair of students, a small group, or perhaps the entire class to show you how they did what they did; how a problem was solved; how a particular manipulative material, technological tool, or related representation was used; and so on. As the examples and Small Group: Show Me Record (Figure 3.2) provided in the previous section of this module indicated, you can use Show Me not only to validate what you have observed but also to provide responses that will help you in considering how you might redirect your instruction within the day's lesson or the planning for tomorrow's lesson.

INSIGHT
Based on previous observations and interviews, you can anticipate the "lesson bumps" where Show Me might be useful.

While it can be paired regularly with student or class observations and serve as a prerequisite or supplement to an interview, the actual use of Show Me will be related to the content focus and student expectations of your lesson. Think about the following questions: "As my students become engaged in this mathematics lesson, what is the range of responses that they may demonstrate? And How can I use Show Me to gain a deeper understanding of their thinking?" As just one example, consider having the students show how they use particular representation tools within the day's lesson (e.g., "Show me how you used the fractions bars for comparing these two fractions. What did you find out?").

Popular considerations for your use of the Show Me technique include showing how particular procedures or representation tools are used. However, and as noted earlier in this module, Show Me is also useful for having students describe their reasoning and demonstrate how they solved a problem. Considering the intent and product of what's to be shown is important as you plan.

Kacey, a second-grade teacher, noted the following:

> *"I use Show Me every day. One of the reasons, not the only reason, I use it is that the student responses provide me with a record of what my kids are doing. I keep copies of their work and use them during parent conferences or, as needed, for chats I might have with my principal about particular students."*

Kacey also noted that her Show Me responses not only provide her with a record of what her students have done, but she can actually see the growth of her students within particular topics across the curriculum.

Martin, a high school algebra, geometry, and precalculus teacher, noted that he uses the Show Me technique a lot because many of his students just need to have the lesson's focus extended a little differently.

> *"I frankly consider Show Me 'hidden practice.' It gives me an opportunity to literally see how they think, how they apply. Much more powerful, in my view, than 'do the odd examples on page x.'"*

Martin is convinced that his students feel better about what they are doing day-to-day and have grown from providing a quick representation and short oral response to problems and tasks of the day. He noted that the use of the Show Me technique helps him define his next steps planning-wise.

An important classroom consideration when using Show Me is to be prepared with regard to particular student responses. Our experience has been that, on occasion, some students could not or would not respond to Show Me or copied the responses of others. While such responses are often unexpected, think about how you might address and monitor these students (e.g., for those unsure about responding to a Show Me request, consider having the student partner with someone and have them show their response together, or consider an interview for such students).

Finally, as you regularly consider particular components of lessons where you would want students to show what they are doing or have done, recognize the value of the performance-based documentation that Show Me provides. But also recognize how particular Show Me responses, not unlike observations, are helpful to you in guiding the pace of your lesson. A Show Me response may suggest a whole-class discussion or review of a lesson topic, but it may also suggest that you quicken the pace of the day's lesson. Perhaps most importantly, Show Me responses identify areas of instructional consideration in the future—for tomorrow's lesson or when you plan to teach similar concepts and skills in future years.

Time to Try

Your use of the Show Me technique will very often be directly connected to what you have observed or the interview comments of your students. Showing is performance-based. A Show Me response is a demonstration of whatever the prompt requests. While your daily planning should anticipate particular aspects of your lesson that would prompt a Show Me (e.g., early learning opportunities using a particular representational tool—number line, pattern blocks, coordinates; solving two-step equations for the first time), Show Me requests also regularly occur based on the serendipity of what you have observed, where the immediacy of a Show Me request and response immediately impacts the flow of your lesson. So, it's now time to try! Let's focus on a mathematics lesson that will occur within the next day or so. Using the following table, enter in brief responses to each of the statements related to planning for and actual use of the Show Me technique. Discuss your "plan" with a colleague.

STATEMENTS	RESPONSES
Mathematics/Lesson Focus	
Anticipation: What are particular components of your lesson that may prompt the use of the Show Me technique?	
Provide several Show Me prompts you may use with a student, a small group of students, or your entire class during your lesson.	
How will you use or adapt the Small Group: Show Me Record (Figure 3.2) for use in your classroom?	
Noting that a student's Show Me response is student-to-teacher feedback, how and when will you provide teacher-to-student feedback to your student, group of students, or class? *How might a student, small-group, or whole-class Show Me response force you to very quickly adjust your teaching of this lesson?*	

Video—Connecting Observations, Interviews, and Show Me

Video 3.7

http://bit.ly/3VmXmEV

Michele, Joanne, and Jessica use the Observations, Interviews, and Show Me techniques and discuss how they connect with each other as they teach and use these techniques to monitor their instruction and student progress. Joanne provides a nice summary of how her use of observations leads to interviews and then perhaps a Show Me prompt. Jessica's fifth graders are engaged in a lesson focusing on volume. Her students demonstrate their student-to-teacher feedback on the Show Me prompts Jessica provides. The teacher comments validate the link between observing, interviewing, and using Show Me.

Think about and discuss how you envision the connections between observing, interviewing, and using Show Me in your own classroom.

SUMMING UP

Not unlike the use of the Interviews technique, daily use of the Show Me technique extends what you observe in the classroom. As you plan, you will consider what you might observe and how such observations may become viable as Interviews and Show Me prompts. These three techniques—Observations, Interviews, and Show Me—are closely connected. They all monitor your lesson's progress and help you consider student and class readiness as you plan for tomorrow's lesson. What's more, they provide anecdotal and performance-based documentation of what your students do mathematically each day. As you plan for observing, interviewing, and using Show Me, recognize that you will first monitor your students via observing as they become engaged in the day's lesson. What you observe may prompt an interview or on-the-spot use of Show Me. Your observations will dictate the frequency and flow of the interview and the extent to which you use both planned or spontaneous Show Me opportunities. Together, the Observations, Interviews, and Show Me techniques serve as that critical filter for differentiation, which is so important to you as you plan and teach mathematics.

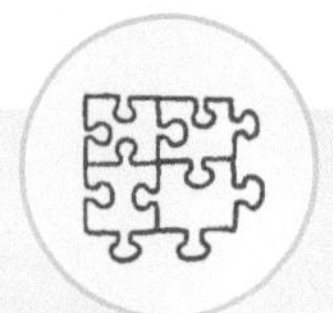

Your Turn

Rate, Read, Reflect! Consider and respond to the following questions with your grade-level or mathematics department teaching team or with teams across multiple grade or course levels.

1. How frequently will you use the Show Me technique?

2. As you consider the use of a Show Me prompt, how will you record student responses?

3. As you teach, you will observe your students, conduct interviews, and use Show Me prompts. These techniques are connected, but what's different about Show Me?

not much; they are similar — student writing a response — student is more engaged

4. What challenges do you envision as you consider daily use of the Show Me technique?
5. Are there particular lessons that you think would more likely engage many more Show Me opportunities than other lessons? Why do you think so? Provide examples of such lessons.
6. How will you use the Show Me assessment technique in your classroom? Provide an example of a Show Me prompt you might use.
7. As you present a Show Me prompt and both see and hear the response, how might you extend the prompt to determine the depth of a student's understanding (e.g., if the prompt is to create fractions equivalent to $\frac{1}{3}$, an extension may be to have the student create fractions equivalent to $\frac{2}{3}$)?
8. What about feedback? As a student responds to a Show Me prompt, how and when will you provide teacher-to-student feedback?
9. Consider the major mathematics topics at a particular grade or course level. Create at least one Show Me prompt for each of these topics. Share the prompts with colleagues, asking them to suggest student Show Me responses, or just try out the prompts with your students.

MODULE 4

HINGE QUESTIONS

"I have learned that my planning must include questions to help me assess where my students are within my lesson. I now do these hinge-point checkups within all my lessons."

—HIGH SCHOOL ALGEBRA, GEOMETRY, AND PRECALCULUS TEACHER

"I seriously think that one of the last things I got 'good' at as a teacher was questioning!"

—SIXTH- AND SEVENTH-GRADE TEACHER

"It took me a while to realize that sometimes I needed to change—while I was actually teaching—the questions that I had planned to ask."

—FOURTH-GRADE TEACHER

"The more confident I feel about my planning, the easier it is to frame questions and then consider responses to help me plan for the next day."

—SECOND-GRADE TEACHER

FROM THE CLASSROOM

Mia and I were going over our school's professional learning (PL) day agenda. We had done the assigned reading for the PL day, but we agreed that we had never heard of a hinge question. So, when we began the day's session on hinge-point and hinge questions, both of us were all ears! One of the big takeaways for me was that a hinge-point question typically occurs when you finish a topic within a particular lesson and are ready to move to the next topic.

But Mia's takeaway was that the hinge question is diagnostic, meaning it's framed to see where the class is regarding the major content focus of the lesson and what students were doing within the lesson. The diagnostic potential of the question has taken both of us, and really most of our teaching colleagues, some time to think about.

We now understand that the hinge-point or hinge question responses really help to define what the class knows and understands about what was taught, which is really important for our planning! We also spent a lot of time just talking about the challenges of questioning, in general, and how it took many of us time to become truly comfortable in planning the questions we were going to use. We agreed that this type of question, the hinge, includes all that and more, since we needed to decide whether or not to use a multiple-choice type of hinge question or a more typically framed question.

Then we spent time assessing student responses to the hinge questions—some of the elementary teachers within our group regularly use every-student-response cards, which they think would work well. Some of our middle and high school teachers wanted to consider providing the hinge question using a digital collaboration tool. At the end of our PL session, Mia and I both decided we need more time on the hinge questions, but we were excited about planning for and using them regularly.

Purpose

As you teach, your instructional intent is for all of your students to be engaged in the learning process, actually doing the mathematics they are learning. Such engagement is dependent on the instructional decisions you make as a teacher. Your use of questioning often launches and frames student engagement. Planning the questions you may use is an important element of lesson preparation. Then what? Student responses to your questions, and your analysis of their responses, define your instructional "next steps." In this module, you will explore hinge questions, including how to plan for their use, and the importance of your analysis of student responses to assess student progress within a lesson and across important mathematics concepts and skills.

Module Goals

As you read and complete activities within this module, you will:

- ✓ Develop and implement a hinge question or hinge-point questions for a mathematics lesson.
- ✓ Reflect on the preparation of and student responses to hinge-point or hinge questions you have created.
- ✓ Review, use, adapt as needed, and reflect on the use and value of the hinge-point tools and the Formative (www.formative.com) technological resource presented in this module.
- ✓ Reflect on how you would provide feedback to students based on their hinge question responses.

HINGE QUESTIONS: BACKGROUND AND BASICS

As noted, classrooms should be places where all students are expected to engage in the mathematics they are learning each day. A significant element of student engagement is found within the classroom discussions that you engineer. Developing the ability to do this, though, is challenging. Being able to process student responses to a question, analyze the thinking expressed by your students, and then guide them through the next instructional steps within a lesson comes with experience. But wait—what about challenges related to actually asking questions, the starting point for any classroom discussion?

You involve and engage your students by asking questions every day. In *Principles to Actions*, the National Council of Teachers of Mathematics (NCTM, 2014) noted the importance of purposeful questions to advance reasoning and sense making, presenting a framework (Figure 4.1) of types of questions important for mathematics teaching and learning. While the question types consider levels of thinking needed for a response, each of the question types is important, depending on the intent of your lesson and the levels of understanding and engagement of your students.

FIGURE 4.1 • **Framework for Questions Used in Mathematics Teaching**

QUESTION TYPE	DESCRIPTION
1. Gathering information	Students recall facts, justifications, or procedures.
2. Probing thinking	Students explain, elaborate, or clarify their thinking, including articulating the steps in solution methods or the completion of a task.
3. Making the mathematics visible	Students discuss mathematical structures and make connections among mathematical ideas and relationships.
4. Encouraging reflection and justification	Students reveal deeper understanding of their reasoning and actions, including making an argument for the validity of their work.

Source: Adapted from National Council of Teachers of Mathematics. (2014). *Principles to actions: Ensuring mathematics success for all.* Reston, VA: Author, pp. 36, 37.

Generally speaking, the questions you provide should be designed to provoke the mathematical reasoning of your students (NCTM, 1991, 2014). Carefully considered and well-posed questions can, and should, both elicit and extend the thinking of your students. As you well know, there are times when student responses to your questions give you a glimpse of student thinking that you had neither expected nor intended. Such responses are important and should clearly influence your planning and instruction.

There are many considerations for you as you continue to develop confidence in crafting questions that are important elements of your teaching and that are helpful to you as you assess student understandings. A first step is to plan the questions you intend to use, considering how you may need to adapt them to address particular student needs, and keeping in mind that the questions you ask are intended to engage your students by causing them to think about the mathematics they are doing. Student responses should help you determine your next instructional steps. Keep in mind that questioning is an integral component of classroom discourse. You orchestrate classroom discourse as you question, listen to student responses, ask students to clarify and justify their responses (orally or in writing), decide when and how to clarify a response, monitor students as they struggle to navigate lesson tasks, and monitor student participation in classroom discussions. Along the way, you must provide time for students to process your questions and consider their responses, and such process time truly varies. Willingham (2009) captured this concern when he noted that "memory is the residue of thought" (p. 54). Your students will need time to process questions posed before they respond. And then, importantly, wait time comes into play. Providing a few seconds (just 3–5!) to wait for a response allows students to think through the question and provide a response that is less tentative. Yes, it's important to provide time for students to "get underneath a question" and decide how to respond. And, yes, you need to provide time to make this happen, but then what? What about the student responses? Smith and Stein (2018) describe five practices for considering and then using student responses in classroom discussions:

1. Anticipating student responses prior to the lesson
2. Monitoring students' work on and engagement with the tasks
3. Selecting particular students to present their mathematical work
4. Sequencing students' responses in a specific order for discussion
5. Connecting different students' responses and connecting the responses to key mathematical ideas

Similar to the Formative 5 assessment techniques presented in this book, these five practices provide helpful considerations as you plan for and guide classroom discussions. Note that anticipating student responses and monitoring student work directly align with the Observations (Module 1), Interviews (Module 2), and Show Me (Module 3) formative assessment techniques as well as the focus of this module—the Hinge Questions technique—while the selection, sequencing, and connecting practices focus on how you might engineer the discussion in your classroom.

Given the everyday importance of classroom discourse generally, and questioning in particular, linking such opportunities to classroom-based formative assessment truly makes sense. The work of Leahy et al. (2005); Wiliam (2011); and Wiliam and Leahy (2015) regarding questioning, diagnostically, at particular hinge points within a lesson allows the teacher to make instructional decisions that impact the pace and instructional sequence of a lesson. The **hinge question** provides a check for understanding or proficiency at a particular hinge point in a lesson. Stated differently, the success of a lesson hinges on responses to such questions as an indication of whether students understand enough to move on.

What About You? Hinge Questions

1. Think about recent lessons you have planned and taught. How often did you ask your class, or groups of students within your class, questions related to the mathematics taught and learned? What was interesting and instructionally helpful about the student responses?

____

____

____

____

2. In your experience with questioning, how do you analyze student responses while teaching and provide feedback to the student responders?

____

____

____

____

3. Think about planning and questioning. As you consider and plan for a mathematics lesson, how have you found time to draft questions you may use, within your lesson, as part of the planning process?

____

____

____

____

INSIGHT

Hinge question responses directly inform both planning and instruction. Your quick analysis of hinge-question responses provides you with in-the-moment signals to guide, monitor, and shift lesson activities as you teach.

4. As you have gained experience in regularly using questioning when teaching, consider and then respond to the following questions related to the five practices for considering and using student responses (Smith & Stein, 2018).

a. How frequently do you draft questions to be used with a lesson, adapting them as needed?

____

____

____

____

b. Do you think anticipating student responses to your question(s) is a helpful planning opportunity? Why?

____

____

____

____

c. How is the use of questioning at particular hinge points within a lesson somewhat of a diagnostic activity?

____

____

____

____

d. How do you determine, as you engage your students in a class discussion of question responses, the sequence of student responses to the questions?

____

____

____

____

e. Describe how you analyzed and then summarized responses to a question you presented in a recent class.

____

____

____

PLANNING FOR USING HINGE QUESTIONS

The hinge question is the question you ask that truly provides feedback to guide your next steps, both within your lesson and as you plan for the next day. Presenting the hinge question typically occurs toward the end of a lesson, but you could ask hinge-point questions at particular points *within* a lesson where you need to get a sense of what students know before moving forward with the lesson. Consider KC's example:

> *"I often have lessons where I am focusing on more than one mathematics concept, so I use hinge-point questions a lot. The responses from my students and my analysis give me a sense of closure for a topic or send a message that my students need more time on this topic before I move on to the next area of mathematical focus that day. Additionally, if I want to do a class check-in during a lesson, just to see how things are going, I use a hinge-point question. When I do this, the student responses and my analysis provide me with an in-the-lesson update of my students' progress and understandings of the focus of today's lesson."*

Let's consider the following connection between the Interviews and Hinge Questions techniques. A hinge question can be thought of as a whole-class or small-group interview, in that the hinge question is diagnostic in focus. As KC's comments note, hinge-point responses indicate what the class or student group knows at that point within the lesson of the day. The hinge question should assess an important element of your lesson, a key concept, or perhaps one of those troublesome bumps on your instructional pathway. In both cases, student responses will help you to define and map your instructional next step or steps. Ideally, your students will respond within one minute, and you will analyze and interpret responses within 15 seconds (Wiliam, 2011). In our work, we consider a two-minute hinge-point window to include posing the question, obtaining student responses, and your analysis of the responses.

Like the Observations, Interviews, and Show Me techniques, the Hinge Questions technique is integral to the planning of a lesson and deciding what is important, both mathematically and pedagogically. Our sense is that the hinge question is that make-or-break query that not only helps you determine your in-the-moment instructional maneuvers but also advises your planning for the next day.

Consider the following different uses and types of hinge questions, one example at the third-grade level and the other with fifth-grade students.

> "Our school's third- and fourth-grade teams have been meeting, probably every other week, to check in on our work with the Formative 5 techniques. At our last meeting we had a lively discussion about hinge-point questions, which included how the student responses helped us in considering our planning for the next day and even adapting, on occasion, the direction of our lessons as they were being taught—based on the hinge responses. My experience with the hinge-point questions has been limited, but I have been pleased with the responses of my third graders. I decided for my lesson, which focused on estimation involving

time in minutes to the nearest hour and then included the addition of two-digit timed amounts and checking for the reasonableness of those results, that it just made sense to use two hinge-point questions. Lots of math going on! After some whole-group work, I used the following:

Maria was competing with her brother, Tad, by keeping track of how long they walked each day. On Monday Maria walked 25 minutes. On Tuesday she walked 30 minutes, and then on Wednesday she walked almost an hour; actually, she walked 55 minutes. She estimated her total walk time for these days to be 2 hours. How long did she walk? Was her estimate a good estimate? Be ready to discuss why.

I had the class use clock models and drawings, if needed. I did a quick sharing of responses. I was pleased, then I moved on to the second stage of my lesson, which was mostly review of activities related to time to the nearest hour and using timed amounts to both estimate and determine the actual number of minutes. Toward the end of the lesson, I used the following hinge-point question to decide just how the students were doing regarding connecting estimation, adding timed amounts (in minutes), and considering reasonableness:

Tad said, "I walked for 25 minutes each day this school week. I know that's a lot more than 2 hours." How long did Tad walk? What can you say about his estimate of the total time he walked?

Like the lesson's first hinge-point question, I had my class use manipulative clocks or drawings as needed to show what they did. Students recorded their answers on their dry-erase boards, so it was pretty easy for me to spot the student timed amounts *and* their work. This seemed to work well, and while I could see their timed amounts for Tad's walk week, I ended up calling on a few of my students for the 'What can you say about his estimate?' part of the problem. I was pleased with the responses, the students demonstrated the correct times, and tomorrow we could move beyond our focus on time to the hour, adding timed amounts and estimated times to expanding our work with adding and subtracting larger numbers. One additional point. I recently started to decide who I will call on with my questions rather than acknowledging the predictable hands that are seemingly always raised (or not)! I think this gives me a better sense of the full range of responses related to what my students know and can do. But the major point here is that I liked using two hinge-point questions, one to assess student progress with the first part of my lesson and the other to serve as my decision maker regarding, in this case, my focus on time to the nearest hour, including adding and estimating timed amounts within a problem context. Both hinge-point questions impacted my work with students today and my planning for tomorrow."

Will and Erin, both fifth-grade teachers, approached their use of the hinge question differently. Here's what Will had to say.

> "Both of our classes were working on the following mathematics standard: interpret division of a whole number by a unit fraction and determine (by computing) the quotients. Erin and I both decided that we would ask our hinge questions toward the end of our lessons. When we first began using hinge questions, we didn't think about providing time for really analyzing student responses and then either adapting our teaching to address any concerns or just moving forward instructionally. We learned that lesson. We now provide more time for the question, our analysis, and any additional instructional support we may need to provide. Both of our end-of-lesson hinge questions were based on the following problem:

Both of the fifth-grade classes were planning a class party and decided to order 12 pizzas. If each fifth grader received $\frac{1}{4}$ of a pizza, how many students would be able to eat pizza?

I used a multiple-choice format for my class, and Erin just used the question with an added phrase. Here's what our hinge questions looked like:

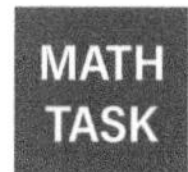

Will's class:
Both of the fifth-grade classes were planning a party and decided to order 12 pizzas. If each fifth-grade class received $\frac{1}{4}$ of a pizza, how many students would be able to eat pizza?

A. 17
B. $\frac{12}{4}$
C. 24
D. 48

Erin's class:
Both of the fifth-grade classes were planning a party and decided to order 12 pizzas. If each fifth grader received $\frac{1}{4}$ of a pizza, how many students would be able to eat pizza? Be prepared to tell me how you know.

Here's the difference in how we framed our hinge questions. I decided to use the multiple-choice format. This took a bit more time to create because I needed to provide the answer choices, and they needed to be choices where I could tell what the students might be thinking when they selected them. For example, if someone selected A, or 17, I was pretty sure they just added the whole numbers in the problem. If they selected B, they probably divided 12 (the pizzas) by 4, essentially determining $\frac{1}{4}$ of the number of pizzas. For each of the choices A–D, I tried to provide a response that was not just random, but one that students might actually consider or do. I like the diagnostic potential of the multiple-choice format of the hinge. I had my students respond by just raising cards—A, B, C, or D—that I provided. This was quick and allowed me to quickly identify error types, and I had a few minutes to consider my next steps. The response to Erin's hinge question was provided orally by the students. She thought that what was most revealing were the 'tell me how you know' responses her students provided. She sampled close to 10 responses, and most were the correct answer. She then moved around the room and did note that some of the students either added or subtracted the 12 and the $\frac{1}{4}$ while maybe three students multiplied $12 \times \frac{1}{4}$. Just like me, Erin had time to do some more with these types of problems before her math class ended.

Both of us agree that there are advantages to both types of hinge questions. Thinking about and then creating the answer choices takes some time in the planning of the multiple-choice-format hinge question, but really does save time in reviewing and analyzing responses. On the other hand, using an actual question without answer choices provides the teacher with an opportunity to hear or see what the students were thinking when they answered the question. Erin and I will both continue to work on hinge questions. We both feel that our hinge question responses really help to guide us in how we will use the time remaining in our mathematics class each day, as well as inform our planning decisions for the next day."

The uses of the hinge question discussed in these examples indicate the level of planning needed to both craft and implement the question. Also note that depending on the focus of the lesson, hinge-point questions might be used to determine particular points of progress within elements of a lesson. Our experience has been that the multiple-choice version of the hinge question is probably used more frequently from the upper elementary grades through middle and high school, and that many teachers like this format as it also serves as practice for seeing and responding to multiple-choice items on end-of-year summative assessments. That said, responses to a hinge question addressing a lesson hot spot, however framed, immediately identify what you will be doing next—within your remaining mathematics time or in tomorrow's lesson. The questions and accompanying statements in Figure 4.2 should be helpful to you as you consider the use of hinge questions in your own classroom. Actual tools for considering and using hinge questions and classroom-based examples are provided in the next section of this module.

FIGURE 4.2 • **Planning for Hinge Questions**

1. How will you use hinge questions as you teach?
• When might you use the multiple-choice format for your hinge question?
• When might you use a more typical question format for your hinge question?
• When might you consider using hinge questions at particular hinge points within a lesson?

(Continued)

(Continued)

2. If you think of the hinge question as a whole-class interview, how will you use the responses?
• When using a hinge question, how will you determine whether your class is ready to move on to the next topic/lesson/standard? • How will hinge question responses influence your on-the-spot, minute-by-minute instructional decision making? • What about feedback? Describe how you will provide teacher-to-student feedback to the class.
3. How will you consider student responses to a hinge question?
• Will students use student response tools (e.g., cards, sticks, devices) when responding to your hinge questions? If so, explain how they will do so. • How will you sample student responses to your hinge question? • How will you ensure that you can efficiently analyze hinge question responses?
4. How will you consider types of questions for your hinge question?
• Think about how you will consider question types—gathering information, probing thinking, making the mathematics visible, and encouraging reflection and justification—when you plan and compose a hinge question. Importantly, how will the mathematics focus of the day determine the type of hinge question you may ask?
5. When you prepare to ask a hinge question, what might you anticipate?
• Have you carefully prepared and thought through possible responses to your hinge question? • Should you ask the hinge question to small groups of students or the whole class? • Should students have access to manipulative materials, dry-erase boards, or devices or other representation tools for their response to the hinge question? • Student responses to a hinge question become their feedback to your prompt (the question). How will you ensure a range of student responses?

As noted, consideration of the statements and questions in Figure 4.2 should be helpful to you as you plan for everyday use of hinge questions both at particular hinge points within a lesson and toward the end of a lesson, depending on the content focus of your mathematics lesson of the day. It's important to remember that the hinge question is diagnostic in that responses should immediately provide an assessment of what your students know about a particular topic. Like the other formative assessment techniques, the Hinge Questions technique will be integral to your lesson planning, and responses to your hinge question will determine your next steps—both in planning and instructionally. The hinge question uses what you do every day with students, which is asking questions and guiding discussion, and directs a particular question to those "I wonder if they are getting this" challenges you think about as you teach. These responses will let you know!

TIME OUT

Let's Reflect:

- As you think about the mathematics lessons you plan for and teach, how frequently do you plan for the questions you use to engage your students and assess their understandings?

____

____

____

____

- Consider a lesson you will plan for and implement in the next day or so. When will you use a hinge question in the lesson? Provide the question.
  - Lesson focus:

____

____

____

____

  - When will the hinge question occur? ______________________________
  - Draft a hinge question:

____

____

____

____

- As you think about creating and implementing hinge-point questions, do you think you would more frequently use the multiple-choice format? Why?

____

____

____

____

(Continued)

(Continued)

- Consider the following comment from Seth, who now "gets it" regarding the use of hinge-point questions:

  "I have been teaching close to 10 years. In all honesty, it took me a while before I became comfortable and confident when asking questions of my students within the lessons I prepared and taught. You could say that I learned from experience and, in fairness, from a helpful observation by my math supervisor. And then I learned about the hinge question! What I do is really think about and plan for the use of a hinge or multiple hinge-point questions within my lesson and then be prepared to modify my questions based on the flow of my lesson. The hinge responses both help me in adapting my lesson 'on the fly' and really help to define my next steps planning- and teaching-wise." What's your takeaway from Seth's comments regarding his use of the hinge question?

  ____

  ____

  ____

  ____

Ponder This:

- Given the importance of questioning as an element of instruction, and the value of hinge-point questions to provide timely classroom-based assessments of student progress within a lesson, offer a brief response as to how you typically use questioning within your mathematics lessons.

  ____

  ____

  ____

  ____

- Based on your reading and the related activities with the Hinge Questions technique, how will *you* use hinge questions? Share a brief comment about how you intend to use the Hinge Questions technique, including how you intend to analyze hinge question responses (student feedback) to the hinge question prompts you provide.

  ____

  ____

  ____

  ____

- Recognizing that when you prepare hinge questions you may generate your question as a typically presented question or as a question with multiple-choice responses, when would you use either type of hinge question format? Provide one example of the mathematics lesson's focus for each of the following question types.

- Hinge question—multiple-choice use:
- Hinge question—typically presented question and response format:

____

____

____

____

Video: Planning for a Hinge Question

Video 4.1

http://bit.ly/3Vz4ePo

Skip Fennell presents a summary of the hinge question and its importance, including the formats of the hinge question. Notice how Sarah, a third-grade teacher, describes her use of the hinge question and Jessica, the school's mathematics coach, discusses planning a lesson and use of the hinge question within the lesson. They then plan a multiple-choice-format hinge question discussing each of the response choices.

Think about and discuss elements of this planning process that you may consider for using hinge questions in your classroom.

TOOLS FOR USING HINGE QUESTIONS IN THE CLASSROOM

As you think about the use of hinge questions in your classroom, reflect on the use of the Observations, Interviews, and Show Me techniques, and consider their connections to the Hinge Question technique. As you observe each day, what you observe sharpens your thinking related to *when* you might ask and *what* you might ask within a hinge question, with what you observe during a lesson having the potential to suggest a revision to a hinge-point or end-of-lesson hinge question you have already planned. As noted, the hinge question is, to an extent, a whole-class or group interview, so the interviews you might plan for and implement, and the student responses to your interview questions, may also impact the planning and use of your hinge question. Consider Angela's and Shondra's very different uses of the Show Me technique and how they connected it to their hinge question.

> "For today's hinge question, I asked my fifth graders if the difference between 7.023 and 7.06 was greater or less than $\frac{1}{2}$. And I also asked them to show me how they knew. For the hinge question response, I had the students raise cards with < or > symbols that I had prepared and provided for student pairs. When I asked for the hinge response, most of

the student pairs were correct. For the Show Me response, I had the student pairs use their dry-erase boards—some of the pairs subtracted the decimal amounts, and some decided to use fraction or number comparisons. I am pleased that I am to the point where I understand the use of the Formative 5 techniques well enough I am comfortable interchanging their use within a lesson."

—ANGELA

"Working with small groups of my fifth-grade class, I asked them to show me how they would determine the product of $\frac{2}{3} \times 9$. As I moved around to the groups to check responses, I noticed that most of the groups used the algorithm for multiplying fractions, but some did use the number line to show two 'hops' of 3 = 6 for $\frac{2}{3}$ of 9. Two of the groups used counters and made 3 groups of 3 counters and then pushed 2 of the groups together to show $\frac{2}{3}$ of the 9 counters, or 6. I like the potential of using my Show Me responses to help guide the development, possible revision, and use of my hinge question. So, based on the Show Me responses, I decided to use a hinge-point question for the whole class, and I asked: 'If we ordered 12 large cakes for the class party and ate $\frac{3}{4}$ of the cakes, how many would we have left?' I intentionally made this a problem the class could do in their heads or work out, but also made it a two-stepper! $\frac{3}{4}$ of 12 = 9, so 12 − 9 = 3 cakes left, which essentially extended the Show Me activity. I had my students raise their hands to respond to the hinge, calling on 9 or 10 students. I wish I had provided multiple-choice responses since several of the students said the answer was 9, while most of those responding did say 3 cakes were left. I really value the opportunity that a hinge-point question provides. In this case, I took time to go over the student responses provided and offer feedback to the class and individual students before continuing with the lesson."

—SHONDRA

Now let's move more directly to using the hinge question by considering the following questions, which are followed by classroom-based responses, and tools for hinge question use.

1. **Think About:** Do I use the hinge question toward the end of every lesson, or can I use it whenever I like?

Classroom Response: *I typically use the hinge question toward the end of my math lesson, but I make sure there is enough time for me to quickly ask the question and review responses (about two minutes total). Then if I need to use the Show Me technique or spend time reviewing a particular response or type of response from the hinge question, I have the time to do so. A few days* *ago, I asked my third graders whether they thought a school day lasted more or less than 9 hours, and many of them thought they were in school more than 9 hours a day. I am not at all sure why so many responded that way, but I had time to challenge their thinking by reviewing with the class when school started—8:00 a.m.—and when it ended—3:00 p.m.—and then asking the hinge question again—suggesting that, if they wanted, they could use drawings, use manipulative clocks, or just consider our classroom's wall clock. Maybe they just needed time to think about my hinge question more carefully, since they sure got it on the second round! This is the main reason I provide some time not just to ask and analyze the hinge question I use, but also for my next steps, instructionally.*

2. **Think About: What's the difference between a hinge-point question and a hinge question?**

Classroom Response: *In my classroom, the hinge question is my daily decision maker—lesson-wise. The responses to the hinge question help me determine what I need to plan for tomorrow's lesson and provide me with a quick sense of what students know—it is diagnostic and opens up the potential for use of the Interviews and Show Me techniques. The hinge-point question is essentially a hinge question that I might use at particular points within a lesson. For example, last week we were working on perimeter and reviewing prior work with addition and subtraction in the same lesson. So, it was perfect for two hinge-point questions, one toward the end of my review work with addition and subtraction. I used the question, "Sandra took flights totaling about 3,500 miles in April and about 2,700 miles in May. How many miles did she travel, and how many more miles did she travel in April than May?" Then toward the end of the lesson, I asked the following hinge-point question related to our work with perimeter: "Can you create a triangle that has a perimeter of 25 centimeters (cm)? Make a drawing to show your triangle and the length of each of the sides on your whiteboards." In both cases, I needed to make sure there was time to analyze responses to the questions, and consider my next steps instructionally, during the time I had left within today's math class and as I planned for tomorrow's mathematics lesson.*

3. **Think About:** Should I use a multiple-choice-format hinge question or just a question?

Classroom Response: *As a district-based and former school-based math leader, I used to wonder about this. At the primary level, particularly Grades K, 1, and 2, I sort of figured that the kids might struggle with multiple-choice responses and weren't taking end-of-year tests using that format, so I just thought, what's the point of that? And then one day at an after-school meeting specifically discussing hinge questions, where I made my pronouncement, Mona took me to task. She said, "You know, in my first grade, every student has response cards, so we can do that. I actually do things like this all the time. I ask a question using a PowerPoint slide and then flash clues, and the students can raise their* T *card if true and their* F *card if false. For example, I can ask, "What can you say about 36?" and then flash the following* *multiple-choice responses one at a time, asking them to raise their* T *card if true and* F *card if false.*

A. *It's almost 60.*
B. *It's > 30.*
C. *It's close to 40.*
D. *It's 1 more than 25.*

Sometimes I add "E. What else?" as a choice. When I do, I always get some interesting responses, but then the hinge takes a bit longer, more than the two minutes I like to use for what we call "hinge time." While I love Mona's idea and it certainly works, I still think the multiple-choice type of hinge question is more for Grades 3–12. However, I do know it takes less time for teachers to pose multiple-choice-format hinge questions; receive responses (electronically, using every-student-response devices, raising hands, or completing a paper-and-pencil version of the question); and analyze them, rather than moving around the classroom and spot-checking written or oral responses to a hinge question. The time demand for the multiple-choice hinge question is on the front end, actually planning the question, and considering the answer choices for a carefully designed multiple-choice hinge. So, should you prefer a focused question and a quick review of responses by several students to the multiple-choice-format hinge question? My experience has shown that, to an extent, the mathematics focus of the lesson, your background and comfort with questioning, and time all play into this decision. The major point is that the hinge question guides your in-the-classroom, in-the-moment next steps and influences your planning that afternoon or evening, as the response to the next question discusses.

4. **Think About:** How and when do I plan for the hinge question?

   Classroom Response: *When I plan a mathematics lesson, my planning includes what I will anticipate observation-wise, elements of my lesson that would cause me to interview students or use Show Me, and when and how I will use a hinge-point question to monitor the progress of my lesson generally as well as monitor those learning bumps that seem to regularly occur in certain lessons. My hinge question just lets me know what my kids know—in the moment. I found the Planning: Hinge Question Considerations Tool (Figure 4.3) to be particularly helpful to me when I first began using hinge questions. It really helped me think about whether my hinge question would be appropriate. Since then, I have "graduated" into everyday use of the Classroom: Hinge Question Implementation Tool (Figure 4.4), which is now my go-to resource.*

FIGURE 4.3 • **Planning: Hinge Question Considerations Tool**

Date: May 8

Hinge Question: Lawnscape Unlimited mowed 12 lawns in their 8-hour workday. At that rate, how many lawns could they cut in a 40-hour work week?

	YES	NO
Will the hinge question assess important mathematical understandings of the day?	✗	
Will students understand the question?	✗	
Will students be able to respond in about a minute?	✗	
Will expected responses be such that they can be analyzed and interpreted quickly?	✗	

General Consideration: Will student responses assist in shaping the planning for tomorrow's lesson?

Circle one: 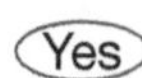 Yes No (If no, revise hinge question)

Yes, and I may use responses to frame an exit task for tomorrow's lesson or have the students create exit tasks we can try out on the class.

How?

We have begun our work on proportional reasoning and rate, extending from prior learning experiences involving fractions my students had at the elementary school level. I wanted to see how they approached the task and whether they computed to determine the problem's solution or perhaps used a ratio table to represent equivalent ratios and the solution. The extent to which ratio tables were used will really influence my planning for tomorrow's lesson.

Source: Fennell, F., Kobett, B., & Wray, J. (2015). Classroom-based formative assessments: Guiding teaching and learning. In C. Suurtamm (Ed.) & A. McDuffie (Series Ed.), *Annual perspectives in mathematics education: Assessment to enhance teaching and learning* (pp. 51–62). Reston, VA: National Council of Teachers of Mathematics. Republished with permission of the National Council of Teachers of Mathematics; permission conveyed through Copyright Clearance Center, Inc.

A blank template version of this figure is available for download at **https://qrs.ly/wsetnnz**

The Planning: Hinge Question Considerations Tool (Figure 4.3) is helpful in early efforts as you consider everyday use of the hinge question, guiding your decision making as you plan for the use of the hinge question. This tool helps you in actually planning for and framing the hinge question. Once you feel more comfortable with the considerations for any hinge question—multiple-choice format or question-and-response format—the Classroom: Hinge Question Implementation Tool (see Figure 4.4) will become an element of your lesson planning. It's most frequently used as teachers implement the hinge question. Download these tools (Figures 4.3 and 4.4) for your own use and adaptation at **https://qrs.ly/wsetnnz**. It should also be noted that the implementation tool (Figure 4.4) could easily be adapted as an individual student Show Me.

FIGURE 4.4 • **Classroom: Hinge Question Implementation Tool**

<table>
<tr><td colspan="3">Date: May 15</td></tr>
<tr><td colspan="3">Mathematics Standard: Grade 7: Expressions and Equations—Solve word problems leading to equations of the form $ax + b = c$ and $a(x + b) = c$, where a, b, and c are specific rational numbers.</td></tr>
<tr><td colspan="3">Hinge Question: The perimeter of an isosceles triangle is 72 cm. If one side is 12 cm, what is the length of each of the two equal sides? Response choices:
A. 24, since $3 \times 24 = 72$ cm
B. 60, since $60 + 12 = 72$ cm
C. 30, since $(2 \times 30) + 12 = 72$ cm
D. 42, since $2 \times 42 - 12 = 72$ cm</td></tr>
<tr><th>LOCATION IN THE LESSON</th><th>ANTICIPATED STUDENT RESPONSES</th><th>POSSIBLE NEXT STEPS: DIFFERENTIATION STRATEGY</th></tr>
<tr><td>Beginning</td><td rowspan="3">Anticipate that most students will recognize the two equal sides with each side being 30 cm. There is some concern that they will not remember/know to respond based on each of the two equal sides, or that some will continue to think rectangles, not triangles. We have worked with rectangles all week.</td><td>Review
Consider reviewing names of triangle types, which may be getting in the way of the standard being addressed in the lesson.</td></tr>
<tr><td>Middle</td><td>Extend
• Use other examples beyond measurement geometry.
• Request equations or expressions within the response (as appropriate).</td></tr>
<tr><td>End
Will use this toward the end of the lesson, using multiple-choice format.</td><td>Student Grouping
Might consider Show Me or an interview based on responses.</td></tr>
</table>

A blank template version of this figure is available for download at **https://qrs.ly/wsetnnz**

As you plan and become acquainted with use of the hinge question, note that we have also found it helpful to try out hinge questions with colleagues in a grade-, department-, and school-level professional learning community. Such rough draft opportunities also provide you with opportunities to both discuss and consider the use of the multiple-choice and question-like hinge question format as well as discuss lessons where hinge-point questions within a lesson, rather than one hinge question toward the end of the lesson, would make sense.

The three examples that follow, one at the primary grade level, one for middle school students, and one for students of high school geometry, demonstrate actual use of hinge questions. What's most important is how the teachers were able to quickly analyze and make the diagnostic decisions that use of the hinge question affords.

Grade 1—Addition and Subtraction: Hannah asked her first-grade students to use their individual dry-erase boards to respond to the following hinge question toward the end of her mathematics lesson:

Bryce has 17 stamps from the United States and Canada. If 6 are from Canada, how many are from the United States?

While she had already planned the content focus (using addition and subtraction within 20 to solve word problems involving addition and subtraction situations) of her small-group instructional rotations for mathematics, she wanted to make sure that her students were placed correctly. She designed the hinge question to ensure accurate placement of the students. She had noticed that student responses to her hinge question often surprised her, and it was critical for her to use this "in time" information to decide the next instructional steps for tomorrow.

Hannah anticipated that many of her students would draw a representation for all the stamps and use some sort of take-from or crossing-out strategy to determine the remaining stamps. She also believed that another group of students would use computation to solve the problem. Her plan was to divide the students into two groups to continue to work on unpacking take-from and compare word problem situations.

As she looked across her classroom at the students' whiteboards, she noted the following responses (Figure 4.5).

FIGURE 4.5 • **Grade 1 Student Work**

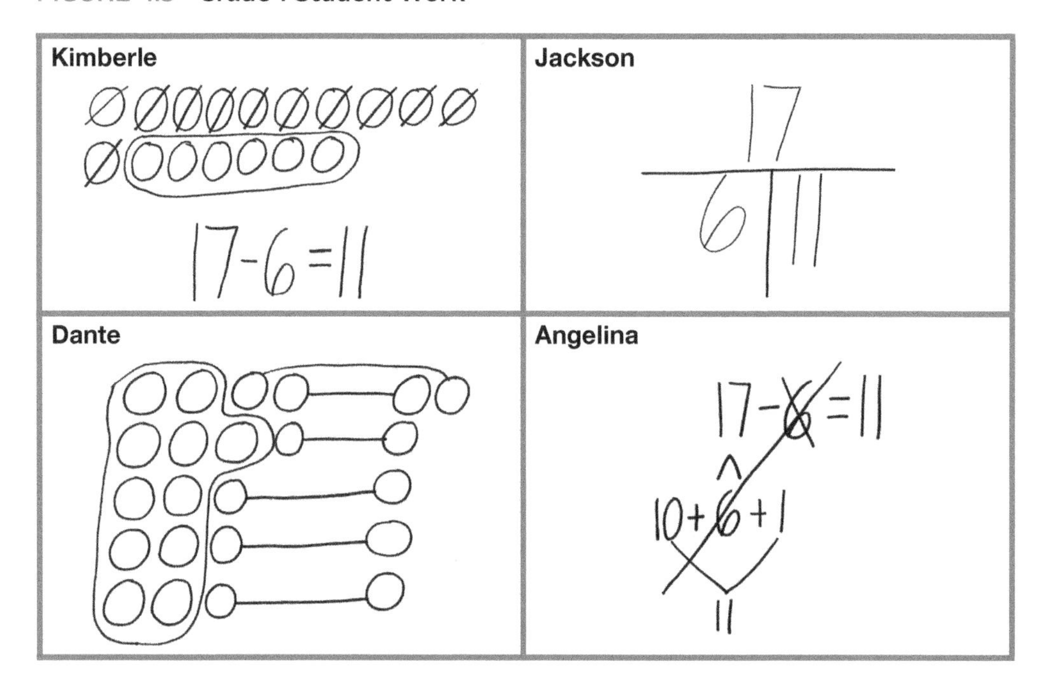

Generally, students responded as Hannah anticipated. Kimberle loved to draw everything to represent her thinking and always seemed to use a cross-out strategy for subtraction. Hannah was glad that Kimberle seemed to have some conceptual understanding, but really wanted to see if she could get her to use some flexible strategies.

Hannah was happy that Jackson used the part-part-whole organizer to show his thinking. She thought he might draw circles and was pleased that he was able to see how this problem fit into the part-part-whole. Hannah was very surprised by two students' responses. As she looked over Dante's response, she noticed that he was using a compare strategy to subtract. Conceptually, this is an important understanding for students, and she is always interested when students record their thinking in this way. She anticipated that Dante would record his thinking with a number sentence and was perplexed that he decided to make a drawing.

Angelina's response was the most surprising. Just last week, Angelina was drawing circles and crossing out items to represent subtraction. Hannah was thrilled to see her decompose to subtract and decided to use her response to begin the small-group instruction discussion.

In summary, Hannah was glad that she used the Hinge Questions technique to guide her instructional grouping. Without this important information, she would have grouped her students incorrectly. Now, she was ready to target the students' instructional needs with purpose and clarity. While this hinge question and the drawings and Hannah's analysis extended beyond her two-minute rule for the hinge question, it was worth it, and she was able to spot responses by this rotation and do her analysis in about three minutes.

Grade 7—Adding and Subtracting Rational Numbers: Melena and her seventh-grade class were studying properties of operations and using them as strategies to add and subtract rational numbers. Today's lesson involved the addition and subtraction of integers. A challenge Melena had been facing in her teaching was how she could seamlessly pose a hinge question, allow students to respond in a timely fashion, and then analyze and interpret the results efficiently. She learned about a digital tool called Formative (www.formative.com) where teachers create/select multiple-choice, short-answer, true/false, or Show Me questions. This tool allows students quick access from any digital device, where they can generate solutions using a digital whiteboard interface. Formative is free, so Melena decided to try it in the beginning of her class one day. She added the following hinge question and sent students the direct link, allowing them to use classroom laptops or their own devices:

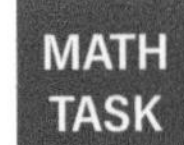

What was the temperature at the end of the day? Samantha was in charge of recording the morning and afternoon temperature for her class. This morning it was really cold. Samantha recorded the temperature as −7°F. The temperature rose 15°F when Samantha had to record the temperature at noon. And then the temperature dropped 6°F when she had to record the temperature at the end of the day. Use a vertical number line to represent the temperature changes and write an equation to represent the changes and temperature at the end of the day.

Melena wanted to see if her students could apply their understanding of addition and subtraction of integers to solve a problem involving rising/falling temperature readings. She also wondered if they would be able to represent the problem and its solution using both a vertical number line and an equation.

Shortly after students began working, Melena was able to monitor their work in real time using the screen shown in Figure 4.6.

She was able to analyze their work quickly and found that most students were able to represent the solution correctly using a vertical number line and an equation. However, upon quick visual analysis, Min's and Rachel's responses included errors, possibly related to interpreting the positive/negative values of certain temperatures, which translated into errors in the vertical number line representations and equations. Melena decided to use the hinge question results to continue with instruction on this topic. She selected Min, Josh, Dan, and Michelle to share their reasoning, allowing them to go up to her computer, select their work, and share with the whole class. Not surprisingly, both Min and Rachel corrected their own mistakes based on comments and questions from their peers. Melena decided to let them both resubmit their work. At one point during the whole-class discussion, Sarita, who created a horizontal number line, revealed that she had never used a vertical number line before. (Who knew?) A classmate reminded Josh that he should write an equation to represent the change in the day's temperature.

FIGURE 4.6 • **"Recording Temperatures" Hinge Question Using Formative Tool (www.formative.com)**

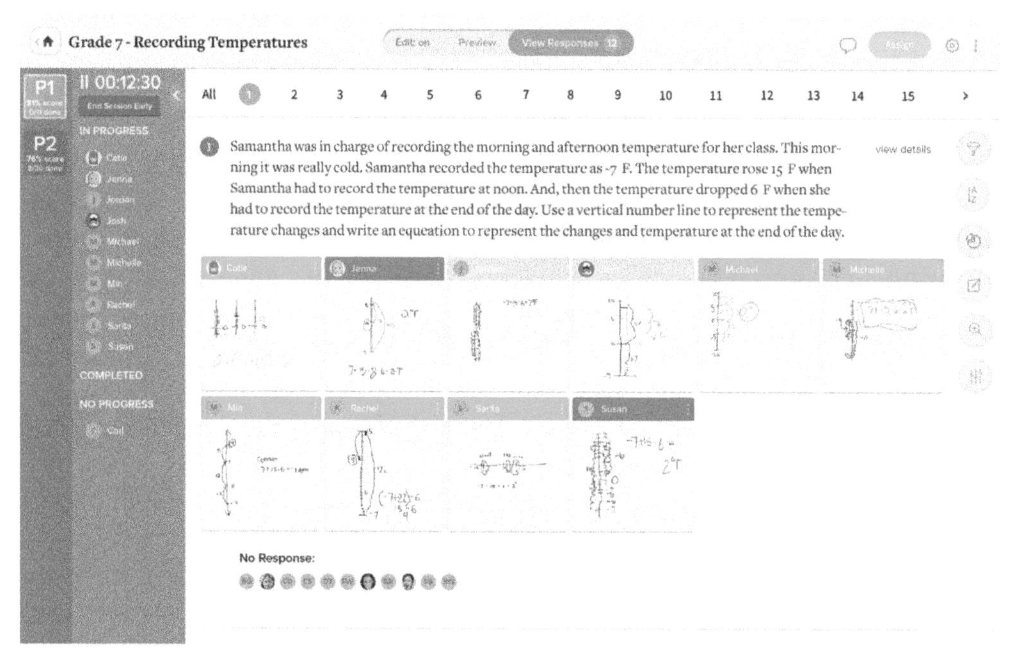

Source: Courtesy of goformative.com.

Melena reflected that in this case, having the digital student learning artifacts available for the entire class to view provided her students with the opportunity to address mistakes or bumps in reasoning, minimizing her role in the process. She also considered that one of the best features of Formative was that it provided a live visual of student responses in action while in the teacher's view, thus allowing her to use the hinge question responses for in-the-moment teaching!

High School Geometry—Congruence and Similarity: The students in Lamar's geometry class have been thinking about the relationship between congruence and similarity. During class investigations, students began to develop an informal understanding of how to use rigid transformations and dilations to show the similarity of any pair of triangles with all pairs of corresponding angles congruent and all pairs of corresponding side lengths proportional. They looked at specific examples and tried to come up with a generalized method that works for any pair of triangles that meets the criteria, much like they did with rigid transformations. They drew on their understanding of the relationship between congruence and rigid transformations and extended that to similarity and dilations. During the cool-down portion of the lesson, students used personal dry-erase boards for each of the hinge question(s) posed by Lamar (adapted from Illustrative Math, Geometry).

Indicate if each statement *must* be true (T), could *possibly* be true (P), or *definitely can't* be true (F). Support your reasoning using a diagram(s).

1. Congruent figures are similar?
2. Similar figures are congruent?

Lamar posed each of the questions in about 20 seconds and gave students about a minute to respond to each statement, while after directing students to hold up their boards, he scanned their work, assessing their responses by looking for the symbol *T* (must be true), *P* (possibly true), or *F* (cannot be true, or false) with accompanying diagrams—in less than 30 seconds.

For hinge question 1, almost all students displayed a *T* (must be true). One student, Holly, wrote *F* (false). Her accompanying diagram (Figure 4.7) displayed what appeared to be two isosceles triangles but with different corresponding side lengths. During Lamar's visual check of student responses, Holly looked around the room, then looked back at her work. "Oh, wait. I drew two similar triangles. They're not congruent," she stated out loud. Holly erased the second triangle and redrew it (Figure 4.8), this time with the same side lengths (and angle measures) as the first triangle. She revised her response to *T*.

When Lamar repeated the process, directing students to display their responses to hinge question 2, their work fell into two categories. Although virtually all students wrote *P* (could possibly be true), the diagrams varied. Some displayed two sets of figures, with one pair similar and with corresponding angles congruent and corresponding side lengths proportional (but with

FIGURE 4.7 • **Holly's Initial Drawing**

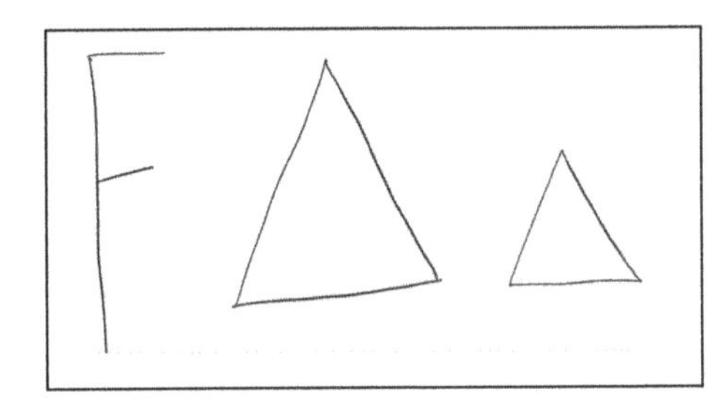

2 similar triangles isosceles "looking"

FIGURE 4.8 • **Holly's Revised Drawing**

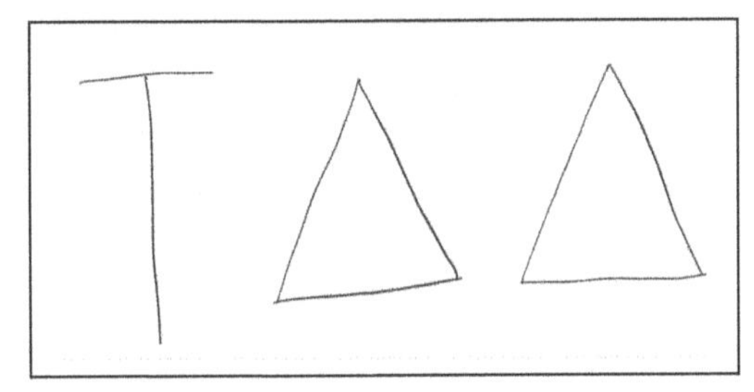

2 congruent and isosceles "looking" triangles—second triangle matches the first

FIGURE 4.9 • **Holly's Drawing With Scale Factor of 1**

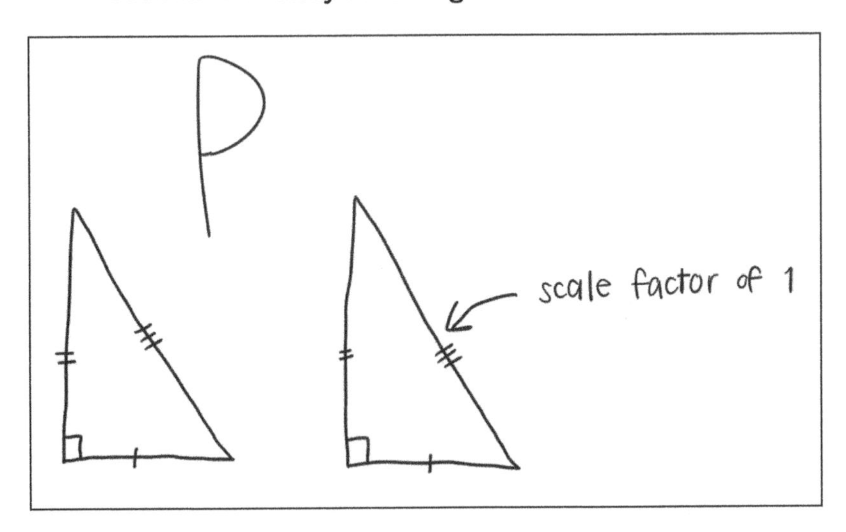

different measures). This time, Holly's drawing showed two identical and congruent figures, with an arrow pointing to the second triangle labeled "scale factor of 1" (Figure 4.9). Lamar concluded that all students were correct and understood. At that moment, he wisely decided to include Holly's and another classmate's responses as part of the next lesson's warm-up, asking, "Do you agree or disagree with each of these responses? Why or why not?"

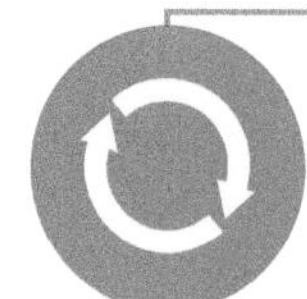

Focusing on Feedback

What about hinge questions and feedback? While hinge questions may typically occur toward the end of a mathematics lesson, responses to hinge-point questions within a lesson provide you with several opportunities to monitor, and even diagnose, student and class progress within a lesson. Like the Interviews and Show Me techniques, hinge questions are essentially prompts with the responses providing feedback to the teacher regarding student understandings of mathematics concepts and skills. While you may consider a general feedback-related comment to individual students as they respond to a hinge question or provide feedback to the class after assessing multiple hinge-point responses, teacher-to-student feedback will be concise. However, taking time at the end of a lesson to write a summary comment regarding class or individual student hinge-point responses is important to not only jump-start your planning for the next day, but provide indicators regarding what you might do differently, instructionally, with the day's lesson, in the future.

Figure 4.10 is presented to show how teacher-to-student feedback can be considered as hinge-point questions are used within a lesson or at the conclusion of a lesson. Note each teacher's analysis of the hinge question responses and, importantly, how they decided to provide feedback to the class or particular students.

FIGURE 4.10 • **Hinge-Point Questions: Questions, Analysis, and Teacher-to-Student Feedback**

Grade Level	First Grade	Fourth Grade	Seventh Grade	High School Algebra
Content Focus	Number and Operations	Geometric Measurement	Proportions	Two-Variable Statistics
Hinge Question	*Use your counters to help solve this problem:* I now have 5 pumpkins. I used to have 3 more. How many pumpkins did I have?	A slide of an analog clock displaying 8:00 is presented to the class. *Which letter provides the best estimate for the angle of 8:00? Be prepared to discuss your response.* A. 90° B. > 90° C. < 90° D. 180°	*A recipe uses 5 cups of flour and 2 cups of sugar. If we want to triple the recipe, how much flour and sugar will be needed?* A. 10 cups of flour and 4 cups of sugar. B. 5 + 3 = 8 cups of flour and 2 + 3 = 5 cups of sugar. C. 15 cups of flour and 6 cups of sugar. D. 21 cups of flour and sugar.	*Consider these two variables: the amount of gas used on a trip, and the number of miles driven on the trip.* *Indicate whether there is:* 1. *a very weak or no relationship* 2. *a strong relationship that is not a causal relationship* 3. *a causal relationship*
Analysis of Responses	I walked around the room, visually inspecting student representations and solutions, then called on several students for their response. Analysis: A large majority of the class "gets this." I am now ready to introduce additional tasks like this one.	Quick show of hands. Almost all selected > 90°. Brief responses from about 8–10 students. Analysis: We are ready to continue our work with angles and angle measurements.	Quick show of hands. Lots picked A and C, but some also picked B and D. I walked by several students to see their work, asking questions about what they were doing. Analysis: I need to work through more example proportion tasks with the class. They need to become more comfortable with doubling and tripling amounts, etc.	Quick show of number using fingers. Most picked the correct answer, 3 (a causal relationship), but some picked 2 (strong, not causal relationship). Analysis: Those who picked answer 2 may be thinking about longer trips (only). I may need to pose a follow-up (assessing) question to students who chose answer 2.
Feedback (Teacher to Student)	General feedback comment to the class that I liked how they represented and solved the problem. Spoke to a few students privately about interesting solution strategies.	General feedback comment to the class about their success with the multiple-choice hinge question. Demonstrated to the class how 9:00 and 3:00 demonstrate right angles (90°).	Feedback to individual students after the class regarding the proportion of cups of flour to cups of sugar.	General feedback (assessing question) to the class: *What impact does the length of a trip have on gas consumption? Explain.* (Longer trips will cause greater gas consumption, and shorter trips will require less gas.)

Consider your class and a forthcoming lesson. Provide a hinge-point question for the lesson and suggest possible student-to-teacher responses as feedback—to you. Then think about and discuss how you might provide feedback to the students, based on their responses. Keep in mind that the suggested student responses to the hinge-point question, their student-to-teacher feedback, and your possible teacher-to-student feedback are just examples, but important rehearsals for your actual use of the hinge-point question in your classroom.

- Grade/Course Level (e.g., Grade 5, High School Geometry):

  ____

- Lesson Focus:

  ____

- Hinge-Point Question:

  ____

- *Suggested* Student Response (Student-to-Teacher Feedback):

  ____

- *Possible* Teacher-to-Student Feedback:

  ____

TECHNOLOGY TIPS AND TOOLS FOR RECORDING HINGE QUESTIONS

There are several online tools that can be used to pose hinge questions and then quickly collect and review student responses. As with all technology tools, be sure that any online applications are approved for use by your school/district and are used in line with existing policies. All of the following tools are designed to respect the privacy of users and are available to use free of charge.

- **Google Forms** (https://docs.google.com/forms): Teachers can develop hinge questions that allow for many response formats (e.g., multiple choice, check all that apply, short answer, extended response, rating scale). Students can use any web-enabled device to read, respond, and even edit their responses (if permitted by the teacher). The teacher can view individual student responses and a summary of student data, analyze data in a spreadsheet, and/or download the spreadsheet.
- **Kahoot!** (https://kahoot.com): Students log in on any web-enabled device with a pin code to respond to a question or questions posed by the teacher. Students receive immediate feedback on whether they are correct or incorrect. A competitive aspect of Kahoot! ranks students on a leader scoreboard based on their accuracy and response speed.
- **Padlet** (https://padlet.com): Padlet is an open space where teachers can pose a question and allow users an opportunity to tap or click on the screen and quickly type a response and even add audio, video, still-image, and document attachment files. After responses have been submitted, the teacher can choose to view the responses in a grid format, which is helpful for quickly assessing student thinking.

- **Pear Deck** (www.peardeck.com): A dashboard provides teachers with a real-time view of student responses, which can be organized, filtered, and managed to help the teacher provide feedback. Aggregate view of responses can help provide the teacher with a glimpse into the thinking of the entire class.
- **Plickers** (https://get.plickers.com): Teachers install this free app and print out the free QR code cards. When setting up a class, each student is assigned a card with a unique number. Each side of the shape on the card corresponds with a different answer choice (A, B, C, or D). After the teacher poses a multiple-choice or true/false question, the students hold up their response cards. The teacher scans the room using the app, which records each student response and displays and archives the data.

Remember that a key to posing hinge questions and assessing student responses includes time-efficient practices. Be sure you test your hinge questions on different devices and browsers ahead of time to ensure their use to support hinge questions is seamless and effective.

INSIGHT

Your experience in the classroom will help you as you anticipate student responses to your hinge questions, as well as potential next steps within the lesson or the next day.

USING HINGE QUESTIONS IN *YOUR* CLASSROOM

As noted, the drafting of hinge questions occurs during the planning of a lesson. Creating and implementing the hinge question will help you in your planning and teaching, as the hinge question will focus on an important component of your lesson. The hinge question should address one of those important instructional hurdles of your lesson. And, if your lesson addresses multiple concepts, skills, or mathematical connections, you may elect to plan for hinge-point questions at the appropriate time (e.g., as the content focus changes) within such a lesson.

Your everyday experience with the Observations, Interviews, and Show Me techniques will actually influence and help you as you prepare for planning and using the Hinge Questions technique. Consider the following comments from Victoria:

> "Hinge-point questions have become an important part of my daily mathematics lessons, which, of course, includes planning for their use. I teach in a small high school in a very rural area. Our mathematics team consists of seven teachers who teach all of the mathematics for Grades 7–12. Several years ago, we began using the Formative 5. My department chair has been just great in helping us all get started in using the five techniques, and she has, in particular, really helped me in my use of hinge-point questions. It took me a while to realize how helpful hinge questions are in monitoring the progress of my students within a lesson. When I talk with our mathematics team, I now tell them about my 'traffic light' use of the hinge question responses of my

> students. Here's what I mean. Sometimes my student responses indicate a red light, meaning I will stop what we are doing and perhaps go over some issues that appear to be challenging to some of my students. Sometimes it's a yellow light response, meaning I may move forward with the lesson the next day or within the period—it's a hinge-point question. But I may also stop or just move gradually with regard to my planning or the rest of the lesson—it's cautionary. And many times, it's a green light response, indicating overall progress and we are moving on. Depending on my lesson, I may use multiple hinge-point questions within the lesson or perhaps use just one hinge question toward the end of my class period. Importantly, thanks to the conversations I have had with our mathematics team, I have learned to provide the time to plan for creating a lesson's hinge-point questions. I often use the multiple-choice format for the hinge question, but I do vary the format of my questions. When I present my hinge-point or end-of-lesson hinge questions, I have learned to provide a 'time cushion' for my selection and analysis of student responses and the in-the-moment next steps guided by the responses. This year our math team has created an online spreadsheet of all our hinge questions. Why didn't we think about this before? Now we can go to our sheet, which we organized by mathematics standard and grade/course level; select a hinge question; and adapt or just reuse it. Sometimes I select a hinge question from the sheet and rewrite a regular-format hinge question into a multiple-choice-format hinge. I have also learned that as I plan a lesson, not only do I think about how my students may respond to my hinge questions, but I also consider how I will monitor their responses—circulating the class, every pupil response, online response, or just sampling student responses. I'm still learning, but classroom-based formative assessments, including hinge questions, are everyday connections to my planning and teaching."

The planning and classroom implementation tools (Figures 4.3 and 4.4) presented in the previous section of the module should be helpful to you as you think about, plan for, and use hinge questions in your classroom. A few considerations may come to mind as you're doing this. Keep in mind that responses to your hinge question are the lesson's "stop and drop" check-in. How are your students doing? What are your next steps? Responses to your hinge questions provide the class pulse you need to have as you consider the pace of your instruction. OK to move on? Do they need more time on a topic? Do you need to interview some students? And finally, the hinge question, together with the other techniques presented in this book, adds to the assessment palette that guides your mathematics planning and instruction and, importantly, provides a comprehensive and living portfolio of the mathematical progress of your students.

Time to Try

Posing effective questions is just one aspect of a larger construct—classroom discourse, which involves a variety of teaching-related actions (Smith & Stein, 2018). Wiliam (2011) notes that there are two good reasons to use questioning in classrooms: to cause student thinking and to provide the teacher with information—student responses—that assists in instructional decision making, both within the lesson being taught and beyond. Knowing about and becoming confident when planning for and using questioning while teaching are related but quite different. Being adept at questioning and using questions to engage is an instructional challenge. As noted, the creation of questions to diagnose what students understand before moving on within a lesson or the next day provides a classroom-based assessment link. The use of such questions at hinge points within your lessons is important, but adjusting to this diagnostic assessment technique will take some time. So, let's get started. It's time to try the technique! Content-wise, we will focus on mathematics lessons you have recently taught or will teach in the next few days. Using the following table, enter in brief responses to each of the statements related to planning for and actual use of the Hinge Questions technique. Discuss your "plan" with a colleague.

STATEMENTS	RESPONSES
Mathematics/Lesson Focus (identify one or two lessons)	
Anticipation: *Do you anticipate using multiple hinge-point questions or a single hinge question? Why?* *Will small groups of students or the whole class respond to your hinge-point or hinge questions? Why?* *As you analyze hinge-point question responses, what will make you stop the lesson you are teaching and return to another topic?* *Similarly, what will make you move more quickly through the lesson you are teaching and perhaps begin a lesson you thought might occur tomorrow or the next day?*	

(*Continued*)

(*Continued*)

STATEMENTS	RESPONSES
Provide a hinge question using the multiple-choice format and also formatted as a typically stated question. *Multiple-choice-format hinge question:* *Regularly stated hinge question:*	
How would you use or adapt the Planning: Hinge Question Consideration Tool (Figure 4.3) as your grade-level, department, or school team begin to plan for and regularly assess drafts of hinge questions?	
Provide an example of a hinge-point question you might use to launch a lesson. *Provide a brief statement as to why you might use a hinge question this way (as a lesson launch).*	

Video—Conducting Hinge Questions

Video 4.2

http://bit.ly/3Upn9Li

This video focuses on actually conducting or using hinge-point questions. Sarah is working with third graders on fourth-grade mathematics standards and prepared to begin work with multiplying a whole number by a fraction, but decided to begin today's mathematics lesson using a hinge-point question to review prior learning related to adding fractions with like denominators (as noted by Jessica, the school's mathematics coach). Sarah used a multiple-choice-format hinge question and responded to several student responses. She quickly assessed that the class understood and correctly responded to her hinge question. She also mentioned her use of sticky notes that she places on student desks to ask certain students to tell her "how they know," essentially moving from a hinge response to a quick interview for these students.

Think about and discuss the use of a hinge-point question as a preassessment of prior learning. Would you consider the use of the hinge question to preassess?

SUMMING UP

The hinge question extends the Formative 5 to include a technique that becomes, to an extent, a diagnostic extension of the interview for a whole class. A hinge question or hinge-point questions are important elements of every lesson. Responses to the hinge question and their analysis essentially identify the starting point for the immediate adjustment of the day's lesson, if needed, and for guiding the planning of the next day's lesson. Depending on the mathematics content focus of your day, a hinge question may be used toward the end of your mathematics lesson or at particular hinge points within your lesson. The hinge question may be framed and presented orally or in writing or as a multiple-choice-format question. Ideally, students will respond within one minute, and you will analyze and interpret responses within 15 seconds (Wiliam, 2011). Our sense is that, optimally, use of the hinge question, student responses, and your analysis takes about two minutes. The hinge question captures a truly important element of the day's lesson and is a key element to your planning and teaching every single day.

Your Turn

Rate, Read, Reflect! Consider and respond to the following questions with your teaching team (grade or course level) or with teams across multiple grade levels.

1. How frequently will you use hinge-point questions?

(Continued)

(Continued)

2. As you consider using hinge-point questions, which format do you think you more frequently use?

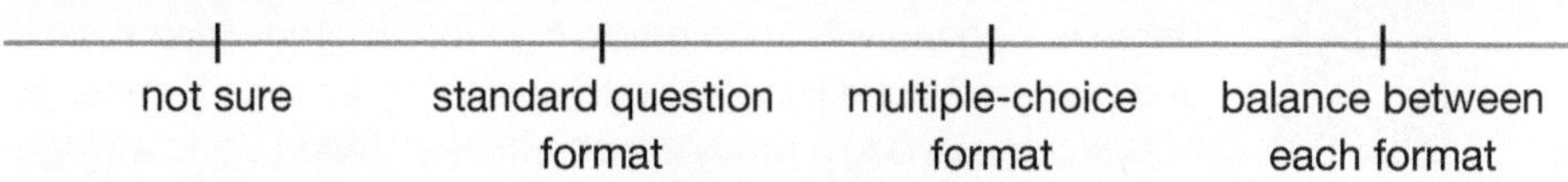

3. As you consider using Hinge Questions as a classroom-based formative assessment technique, what concerns you the *most* about planning for and using hinge questions?
   a. Actually, making sure my questions are good—appropriate.
   b. Deciding whether to use the multiple-choice or standard question format.
   c. Being able to engage my students in responding to the hinge-point or end-of-lesson hinge question.
   d. Doing the on-the-spot analysis of the student responses.
   e. Considering how and when to provide teacher-to-student feedback related to the hinge question response.
4. The hinge question is, to an extent, a diagnostic assessment of class progress. How does this differ from your current way of questioning students within a lesson?
5. What can you learn from your use of the Observations, Interviews, and Show Me techniques to assist you in your use of the Hinge Questions technique?
6. Name a particular mathematics content topic you teach that may be more appropriate for the multiple-choice format of the hinge question. Why was this topic selected?
7. What technological tools might you use for student responses to your hinge questions, and why? Feel free to present technological tools other than those presented in this module (see "Technology Tips and Tools for Recording Hinge Questions").
8. What about feedback? As students respond to a hinge-point question within a lesson, how and when will you provide teacher-to-student feedback to individual students or perhaps a class?
9. Planning for and creating hinge questions takes time. How could you and other members of your teaching team (grade level, department, or school) plan for and develop draft hinge questions, implement them, and then refine them to create a "living resource" you could visit and use on a regular basis? Discuss how such a plan could work.
10. Considering what you have learned about hinge questions and their use, what would you suggest to a student teacher or early career teacher about creating and using hinge questions?

NOTES

MODULE 5

EXIT TASKS

"It really has taken me a while to move beyond what we used to do for an exit ticket to using an exit task. I get it; my exit tickets were trivial—just quick, mostly routine math computations. Now my tasks focus on problem solving and reasoning."

—THIRD-GRADE TEACHER

"I admit to struggling when trying, particularly at the end of a long day, to create exit tasks. I now have some great sources, which I use regularly to jump-start my thinking as I create or adapt previously used tasks. How my students do on their exit tasks is very important to me. Their progress helps define my next steps."

—SIXTH-GRADE TEACHER

"I actually start all of my classes with a task, which pretty much 'drives' my lesson. But my work with the Formative 5 has helped me devise the following. Several days of the week—usually Monday and Wednesday, and sometimes Friday too—I will have a second part of the lesson's opening task, which I use as my exit task. This has worked great—my exit task connects with the lesson and with the opening task, and my review of the student responses helps me regarding short-term planning."

—HIGH SCHOOL MATHEMATICS TEACHER
(ALGEBRA AND GEOMETRY)

FROM THE CLASSROOM

Reggie, Charise, and I were all paired up for our after-school professional learning (PL) session on exit tasks. The biggest question we all had was: What was the difference between an exit task and an exit ticket or slip? Angela, our school's math coach, engaged our learning community in an activity where we did an online search for exit tickets and exit slips, and then we analyzed the mathematics assessed by the tickets and slips. We found that the tickets and slips assessed pretty low-level stuff.

Angela's point was that our end-of-lesson exit task should be a task that challenges and that really engages our students into digging into the mathematics they are learning. We got that and really liked the online activity, but the whole group also discussed not only where to find or adapt tasks for use as exit tasks, but how frequently they could or should be used.

Some of our teachers, I am sure, will provide an exit task each day. Others will do so just a few days of the week, because of the time needed to review task responses and provide feedback to the students.

One of our new teachers raised a really good point. She thought that each of the grades should develop and store exit tasks electronically, so we can use them across grade levels and adapt them as needed. I loved that idea, as did everyone.

At our next meeting Angela is going to help us set up a template and online organization system for the exit tasks. This was a really productive PL session. It ran over time, but everyone stayed and really seemed involved in what we were doing.

Purpose

Exit tasks can be regularly used as classroom-based formative assessments to provide a perspective of a student's or class's understanding of the major focus of a lesson. They can be administered at the end of a lesson or at the end of the day. In this module, you will have an opportunity to view exit (capstone) tasks that can capture the focus of your mathematics lesson, and explore how they provide a sampling of your students' performance and understanding.

Module Goals

As you read and complete activities within this module, you will:

- ✓ Reflect on the importance of an exit task as a critical component of any lesson's closure.
- ✓ Create exit tasks and analyze and provide teacher-to-student feedback based on the responses to the exit tasks.

✓ Identify resources that may be valuable when creating exit tasks or adapting problems/activities into exit tasks.

✓ Plan for the implementation of exit tasks.

EXIT TASKS: BACKGROUND AND BASICS

"Effective formative assessment involves using tasks that elicit evidence of students' learning, then using that evidence to inform subsequent instruction" (National Council of Teachers of Mathematics [NCTM], 2014, p. 95). This module presents Exit Tasks, our fifth and final classroom-based formative assessment technique. We consider the exit task as an end-of-lesson formative assessment. Note that the emphasis here, language-wise, is on an exit *task* rather than the more commonly used exit *ticket* (or slip). The intent of the entry (or exit) ticket (or slip) strategy (Fisher & Frey, 2004) is to help students summarize and reflect on their learning.

INSIGHT
*An **exit task** is a capstone problem or task that captures the major focus of the lesson for the day or perhaps the past several days. The focus and use of the exit task is assessment.*

An **exit task** is a capstone problem or task that captures the major focus of the lesson for the day or perhaps the past several days. The response provides a sampling of student performance. Why an exit task? Many teachers recognize the instructional importance and value of mathematics tasks that truly engage their students, understanding that the way students think about solving problems is governed by the mathematics tasks assigned (Kobett et al., 2021). So, mathematics tasks are regularly used to open a lesson, perhaps as a warm-up activity focusing on a prior lesson, or to "drive" the instructional focus of the lesson. The focus and use of the exit task is assessment.

For many, locating or creating a mathematical task to be used as an exit task can be a challenge. In fact, it's a multifaceted challenge. Not all tasks provide the same level of opportunity for student thinking and learning (Hiebert et al., 1997; Stein et al., 2009). We know that mathematical activities can range from an exercise or set of exercises to a complex problem or task that involves much higher cognitive demand. Consider the following two examples:

Exercise	High Cognitive Task
$\frac{3}{8} \times 4 =$	Use a drawing or manipulatives to represent the solution to the following problem: The class shared 4 cakes at the party. When they cleaned up after the party, the students found that $\frac{3}{8}$ of each cake was left over. How much cake was actually left? More than one cake? Less than one cake? Show how you know.

Smith and Stein (1998) have developed a Mathematics Task Analysis Guide (Figure 5.1) that provides characteristics of low-level and high-level tasks and can be used to analyze tasks that you may select or create for use as exit tasks.

FIGURE 5.1 • **Mathematics Task Analysis Guide**

LEVELS OF DEMANDS
Lower-Level Demands (Memorization)
• Involve either reproducing previously learned facts, rules, formulas, or definitions or committing facts, rules, formulas, or definitions to memory. • Cannot be solved using procedures because a procedure does not exist or because the time frame in which the task is being completed is too short to use a procedure. • Are not ambiguous. Such tasks involve the exact reproduction of previously seen material, and what is best reproduced is clearly and directly stated. • Have no connection to the concepts or meanings that underlie the facts, rules, formulas, or definitions being learned or reproduced.
Lower-Level Demands (Procedures Without Connections)
• Are algorithmic. Use of the procedure either is specifically called for or is evident from prior instruction, experience, or placement of the task. • Require limited cognitive demand for successful completion. Little ambiguity exists about what needs to be done and how to do it. • Have no connection to the concepts or meanings that underlie the procedure being used. • Are focused on producing correct answers instead of on developing mathematical understanding. • Require no explanations or explanations that focus solely on describing the procedure that was used.
High-Level Demands (Procedures With Connections)
• Focus students' attention on the use of procedures for the purpose of developing deeper levels of understanding of mathematical concepts and ideas. • Suggest explicitly or implicitly pathways to follow that are broad general procedures that have close connections to underlying conceptual ideas as opposed to narrow algorithms that are opaque with respect to underlying concepts. • Usually are represented in multiple ways, such as visual diagrams, manipulatives, symbols, and problem situations. Making connections among multiple representations helps develop meaning. • Require some degree of cognitive effort. Although general procedures may be followed, they cannot be followed mindlessly. Students need to engage in conceptual ideas that underlie the procedures to complete the task successfully and that develop understanding.

(Continued)

(Continued)

Higher-Level Demands (Doing Mathematics)
• Require complex and nonalgorithmic thinking—a predictable, well-rehearsed approach or pathway is not explicitly suggested by the task, task instructions, or a worked-out example. • Require students to explore and understand the nature of mathematical concepts, processes, or relationships. • Demand self-monitoring or self-regulation of one's own cognitive processes. • Require students to access relevant knowledge and experiences and make appropriate use of them in working through the task. • Require students to analyze the task and actively examine task constraints that may limit possible solution strategies and solutions. • Require considerable cognitive effort and may involve some level of anxiety for the student because of the unpredictable nature of the solution process required.

Source: These characteristics are derived from the work of Doyle on academic tasks (1988) and Resnick on high-level-thinking skills (1987), the *Professional Standards for Teaching Mathematics* (NCTM, 1991), and the examination and categorization of hundreds of tasks used in QUASAR classrooms (Stein, Grover, & Henningsen, 1996; Stein, Lane, & Silver, 1996). Republished with permission of the National Council of Teachers of Mathematics, from Smith & Stein, 1998; permission conveyed through Copyright Clearance Center, Inc.

Examples of the task types include the following:

Lower-Level Demands (Memorization)

- Primary (K–2) Level:

  $4 + 5 = ?$

- Intermediate/Middle (3–8) Level:

  What is the formula for the area of a triangle?

- High School Algebra:

  What is the expression for the phrase "29 more than a number *n*"?

Lower-Level Demands (Procedures Without Connections)

- Primary (K–2) Level:

  $32 + 15 = ?$

- Intermediate/Middle (3–8) Level:

  $\frac{8}{10} \div 2 = ?$

- High School Algebra:

  Simplify: $(11 \cdot t) \cdot 12$

Higher-Level Demands (Procedures With Connections)

- Primary (K–2) Level:

  Use place value blocks to represent the following comparison question: Which is greater, 64 or 53? Show how you know.

- Intermediate/Middle (3–8) Level:

Use the number line to show $\frac{1}{2} - \frac{1}{8} = ?$

- High School Statistics:

Using the scores, create a boxplot showing the quartiles and median scores and highest and lowest scores—outliers.

Higher-Level Demands (Doing Mathematics)

- Primary (K–2) Level:

Freddie and his dad rode bikes from their house to the high school. The bike ride from Freddie's house to the high school was 1 mile long. Then they rode from the high school to the park, and that ride was 2 more miles. Freddie and his dad had so much fun that they decided to do the same ride the next day. How many miles would they have ridden at the end of both days of bike riding? Be prepared to show how you solved this problem.

- Intermediate/Middle (3–8) Level:

Solve the following problem: The class paid \$80 for 20 hamburgers. If they needed to double their order (from 20 to 40 hamburgers), what would be the cost per hamburger? Be prepared to discuss your solution to this problem.

- High School Algebra:

Here are three patterns of dots.

Pattern A

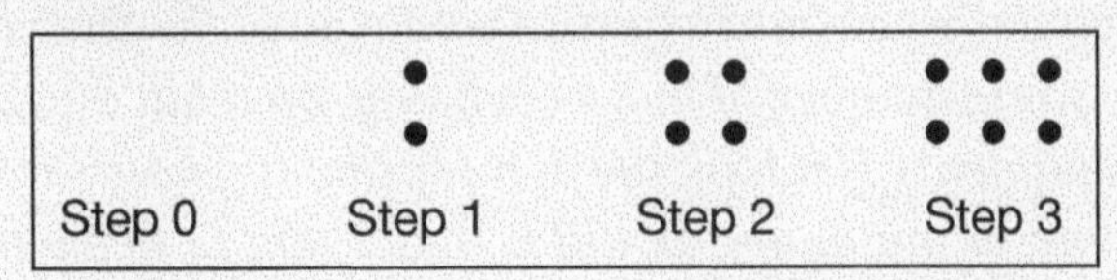

Pattern B

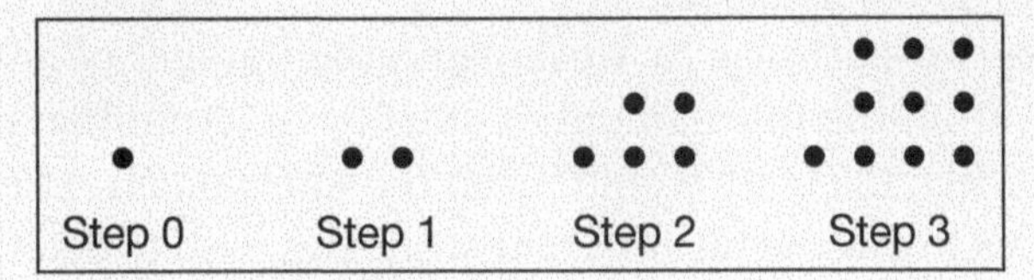

Pattern C

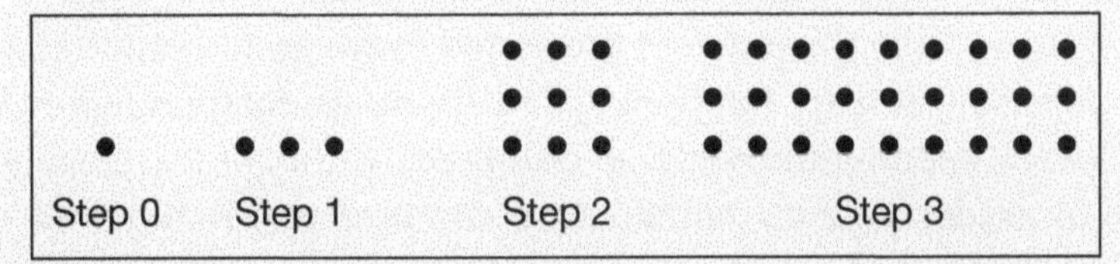

Which pattern shows a quadratic relationship between the step number and the number of dots? Explain or show how you know.

Webb's (1997) Depth of Knowledge (DOK) model is often used to analyze the cognitive demand of assessment tasks. The DOK levels are as follows:

1. Recall and Reproduction
2. Skills and Concepts
3. Short-Term Strategic Thinking
4. Extended Thinking

The DOK levels, not unlike the levels of demand presented in the Mathematics Task Analysis Guide (Figure 5.1), are helpful as you consider the level of cognitive demand of the mathematical tasks you locate, repurpose, or create for your students.

INSIGHT
The actual focus of your lesson very much dictates the level of task that you will use for your exit task.

The actual focus of your lesson very much dictates the level of task that you will use for your exit task. There are times when an exercise that might be considered low-level—or Recall and Reproduction, if you prefer using the DOK levels—is an appropriate exit task capstone to a lesson. Our hope is that all lessons engage students in doing mathematics, thus promoting regular use of problem-based, high-level or thinking-related mathematics tasks as your exit tasks.

Spangler et al. (2014) discuss the potential of the "open-middle" task. One could suggest that such tasks are a variation of an open-ended task—one in which there are many possible solutions with seemingly endless routes to a solution. In contrast, the open-middle task has one correct solution, but the task, as presented, allows multiple paths to the solution. Consider the following:

Cora and her soccer team were driving to an end-of-summer-season soccer tournament. Cora's team had to drive about 500 miles to the tournament. If they drove about 175 miles on Friday, the first travel day, and about the same number of miles each day, on what day would they get to the tournament? If it took the drivers about 3 hours to drive 150 miles, about how many hours would it take them to drive to and from the tournament? Be prepared to discuss your reasoning.

Students can enter this two-part task by working on the number of hours it would take to drive to and from the tournament or start by determining, based on their Friday departure, when the team would reach the tournament. Students will have many opportunities to discuss the different strategies used to determine when the team would arrive at the tournament and a reasonable estimate for the number of hours driven to the tournament and then back home.

Open-middle tasks, like the one just presented, have great potential as exit tasks. Nanette Johnson, Robert Kaplinsky, and Bryan Anderson created Open Middle (https://www.openmiddle.com), an online repository of K–12 open-middle tasks tagged by grade, domain, mathematics standard, and DOK level.

Knowing that there is one, just one, solution to an exit task allows you to spend time considering the varied routes or paths your students used to arrive at their solution. This is not to imply that you should not use open-ended tasks, but issues related to efficiency regarding the use and efficiency of exit tasks must be considered.

The exit task is the formative assessment closer of your day. It provides a demonstration or documentation of student performance on a task of particular importance for the day's lesson. Consider the connection between the Show Me and Exit Tasks techniques. Both provide a performance-based documentation of what your students can do mathematically. While Show Me is used more frequently with individual students or small groups of students, the exit task is typically for all students, although you could differentiate tasks for particular students or certainly suggest that groups of students work together to solve the task.

What About You? Exit Tasks

- While you may regularly use mathematics tasks to introduce a lesson and perhaps serve as the focus of a lesson, how do you locate such tasks? Do you create them? Do you use a textbook or related resource for such tasks? Discuss typical sources that you like to use and how and when you create your own tasks.

  ____

  ____

  ____

  ____

- How do you use the work of your students as they solve mathematics tasks to also engage them in providing student-to-student feedback regarding the solution strategies used to solve a task?

  ____

  ____

  ____

  ____

(Continued)

(*Continued*)

- React to the following statement from a teacher who assigns exit tasks to small groups of students and has them use varied representations to assist in the task's solution. Note how the use of and engaging the representations assists this teacher's multilingual students in communicating.

  "I have a diverse class, many of whom are multilingual learners, so for my exit tasks, I have my students work in small groups and use varied representations as they work through the exit task of the day. These representations include, as appropriate, tables, graphs, drawings, equations, or just words. I then make my rounds of the groups, asking for explanations of the task's solutions and having the students draw connections between all the representations they use. This way, I can essentially see that all students have been engaged in the task's solution, and use of the varied representations seems to really help my multilingual students in sharing their thinking. Importantly, I can provide my feedback to individual students and the group as I make my visit."

  ____

  ____

  ____

  ____

- How should your analysis of the exit tasks of your students help you differentiate your planning and instruction to meet the needs of your students? Share your response with others.

  ____

  ____

  ____

  ____

- Because of the time it may take you to review the exit task responses from your students, you may decide to use an exit task every other day or within certain lessons. Think about your teaching. Suggest a recent or upcoming lesson or series of lessons in which you would create or otherwise locate and use an exit task. Describe the lesson or lessons and provide brief notes about the proposed exit task in the space provided. Discuss your response with other members of your grade-level or department-level team.

FOCUS OF THE MATHEMATICS LESSON OR LESSONS	EXIT TASK NOTES

- **Video:** Plan for and video a student or small group of students working on an exit task you have created or adapted for use in your classroom. Play the video back and provide a statement summarizing student progress with the exit task, including any concerns you may have noted as you watched and listened to the video. Share your summary statement with a grade-level or department-level colleague.

____

____

____

____

Ponder This:

- As noted, you may often begin a lesson with a task. What if you decided to use an exit task as an extension or next step to the task you use to introduce or provide a context within a lesson? How would you propose doing this?

____

____

____

____

PLANNING FOR USING EXIT TASKS

The exit task provides an actual documented sampling of student performance. It provides a record for you to monitor particular student, small-group, and class performance. Coupled together, the hinge question and exit task bring mathematical closure to your lesson and provide the seeds for planning and instruction related to tomorrow's and future lessons. You will plan for use of a particular exit task prior to its implementation, which will occur toward the end of the day's mathematics lesson. Different from the hinge-point questions, exit tasks will require more response time for students to demonstrate their understandings. The purpose of the exit task is two-fold. First, you will provide students with an opportunity to demonstrate what they know and understand, which will help you plan and design your next steps instructionally. Second, you may use your analysis of the exit task responses to provide explicit feedback to your students. As you create and select these tasks, you may also want to think about how you will provide student-to-teacher feedback. Consider the following commentary involving Claudia and Darshan and their use of the exit task.

> "Claudia and I teach in the same K–8 school, but she teaches eighth grade and I teach kindergarten. I use an exit task with my kindergartners at the end of math time almost every day. As I plan my next day's lesson, I think about where my class should be at the end of my lesson and then 'get' a task that I think is appropriate. I get a task by going to my files or to websites I particularly like. For example, some of the sites I routinely visit include the following:
>
> - **NCTM Illuminations:** https://illuminations.nctm.org
> - **Inside Mathematics:** www.insidemathematics.org/performance-assessment-tasks
> - **Illustrative Mathematics:** https://tasks.illustrativemathematics.org/content-standards
> - **YouCubed:** www.youcubed.org/tasks
>
> I remember picking up a task at a school district workshop using digit or number cards. I knew that I wanted to open up the task to encourage a wider range of student responses. I showed Claudia the original task and my revision:

Original Task

Alane has the following number cards: 4, 9, and 12.

- *Make a set showing each number.*
- *Which set has the most? The least?*

Revised Task

Alane has the following number cards: 4, 9, and 12.

- *Show each number using counters or a drawing.*
- *With counters or a drawing, make a new group that is between 9 and 12.*
- *Using a number line from 0 to 20, order the numbers on a number line.*
- *If Alane added 3 to each group, what would the new numbers be?*
- *Would adding 3 to each number change the order on the number line?*

I like to use the sites noted because they can really jump-start my thinking. Our district also has links to other great sources. Oh, and I often just create my own tasks. After adapting tasks for a couple of years, I've gotten pretty good at creating my own exit tasks."

—DARSHAN

"Darshan's kindergarten teaching team and my eighth-grade team were paired together at our last faculty meeting when we talked about the Formative 5. This was funny, in a way, since we represent the bookends of our K–8 school! We decided to talk about exit tasks, and I listened to what Darshan said, but of course, I have my own take. Like Darshan, I often use sites I have found or that the school district provides to locate tasks. Like Darshan, I tend to adapt a task to fit my needs. It's funny. We both agreed that sometimes after our teaching day when we do our planning for the next day, it's much more likely that we go to a site we trust and locate a task online and download it to jump-start our exit task development. In talking about all this, we agreed on one major question and four criteria for tasks we select:

Does this task connect directly to my plans for the day?

1. Is the mathematics appropriate?
2. Does the task engage mathematical practices or processes important to this lesson?
3. Is the level of learning/engagement of the task—low or high or DOK 1–4—appropriate?
4. Will I be able to adapt the task to help me engage my students as full participants when doing "this" mathematics?

In our eighth-grade and kindergarten-level discussions, it was also interesting to note that both grade levels liked the same online sources for tasks, with each of us having other favorite sites. One difference, though. Our tasks at the eighth-grade level tend to be pretty involved—almost always high-level tasks—so I don't use the exit task every day. I use an exit task, for the most part, three days a week—Monday, Wednesday, and Friday. That gives me time to read, review, provide feedback to the students, and plan my next steps."

—CLAUDIA

"I was really glad that Claudia brought up the point about particular websites she and others like—actually, the word should be *trust*. This has become a bit of a sticky issue in our team. One of our members regularly uses a website where she pays for the tasks she uses. We finally pointed out that a lot of what she was bringing to us were essentially fancy worksheets. The mathematics tasks seemed like fun but didn't match the mathematics standard indicated and frankly were just exercises. We finally decided we would only gather tasks from the sites that were reputable and indicated support from experts in the field of mathematics education."

—DARSHAN

Darshan and Claudia's discussion of their use of the exit task addresses many of the everyday questions about the use of the exit task that you will confront. As you plan for your use of the Observations, Interviews, Show Me, Hinge Questions, and Exit Tasks techniques, one important difference is that there are many online and print resources that you will be able to consider as resources for exit tasks. The questions provided in Figure 5.2 should help guide your thinking regarding selection, creation, and use of the exit task as a classroom-based formative assessment technique.

Responses to the statements and questions provided in Figure 5.2 should also help guide your thinking about the use of the exit task and the resources that may be helpful to you in creating or adapting tasks for classroom use. Important considerations to keep in mind include the following:

- The exit task is typically the final activity assessment-wise in the lesson; to an extent, it's a whole-class problem-based Show Me activity.
- You should provide time for all students to complete the exit task, since this is *not* intended to be a speeded activity.
- If you complete a particularly important topic during your mathematics lesson, you may want to consider presenting the exit task then, rather than waiting until the end of your mathematics time. This decision is mostly related to the pace of your teaching and your own comfort, and that of your students, in responding to exit tasks.
- The exit task assesses important mathematics and is often a high-level task (see the earlier discussion) that intersects important mathematics and particular mathematical practices (National Governors Association Center for Best Practices & Council of Chief State School Officers, 2010) or processes (NCTM, 2000).

- You may not want to use an exit task each day, since the task responses will need to be analyzed and considered for your next steps planning-wise.
- Exit task responses provide a record of student performance and can be kept for parent conferences and longer-term analysis.
- Your grade-level team, mathematics department, or school-based learning community may consider organizing and regularly updating exit tasks for use within various grade levels and courses.

FIGURE 5.2 • **Planning for the Exit Task**

1. How will you consider using an exit task?
• The exit task provides an end-of-lesson record of individual student performance. How will you analyze the responses of individual students, small groups, and the entire class? • How will you use the analysis of exit task responses in considering particular student or class differentiation decisions?
2. Where do I get these exit tasks?

• Does your grade-level or mathematics department learning community or school have particular resources that you use for mathematics tasks?
    • What are your favorite sites and resources for mathematics tasks?
• What is your comfort level with creating your own exit tasks or adapting tasks you have located online or from print resources you may have?
• Consider the following criteria when you select tasks, particularly online or previously published tasks:
    • Is the mathematics appropriate as an exit task for this lesson? Does it match what I have taught or will teach?
    • How will the task engage my students?
    • Is the task *appropriately* challenging (see the discussion of levels of tasks and the DOK provided in the previous section of this module)?
    • Will I be able to use the results to guide my planning and teaching, and use the results as a component of my ongoing record of student progress?

3. How will you plan for use of the exit task?
• As you plan for the day, what mathematics will be assessed using an exit task? • Will the task be a lower-level or DOK 1–2 task, or a higher-level or DOK 3–4 task? • How will you check for the mathematics content and mathematical practices or processes being addressed in the exit task? • During mathematics time, when will your students complete the exit task?   • If you complete an important component of your mathematics lesson prior to the end of mathematics time, will you use your exit task then or wait until the end of mathematics time? • How will you use exit task results in your own planning for future lessons?

(Continued)

(Continued)

4. Given your class needs and the value of the exit task, is this a formative assessment technique that you will use every day? Several days a week?

5. How will you summarize student responses and provide feedback for your exit tasks?

- Do you have record-keeping tools that you have found to be helpful to collect or summarize student responses to mathematics tasks?
- How do you plan to provide comments to students as feedback on their exit task solutions and related responses?
- When will you provide feedback to your students on their exit tasks?
- How much time will you need to discuss the feedback with your students? When will you do this?
- How will you endeavor to provide feedback that focuses on what your students know, using such strength spotting to leverage a discussion related to their challenges?

6. How will you organize your classroom to implement the exit task?

- How much time will your students need to complete the exit task?
- What resources (e.g., manipulative materials, drawings, paper, online access) will be needed by students completing the exit task?
- Are there any special seating arrangements needed for use of the exit task?

TIME OUT

Let's Reflect:

- As you plan for regular use of the exit task, what are your most pressing concerns related to creating or acquiring such tasks and adapting them to meet the needs of your students?

  ____

  ____

  ____

  ____

- The exit tasks you provide should, of course, address important mathematics standards and mathematical practices or processes. Your tasks should also challenge and engage your students. Jansen (2020) suggests that such tasks invite "rough draft thinking" by engaging your students in representing and discussing varied solution paths and solutions, helping to ensure a collaborative and inclusive classroom mathematics environment that invites student-to-student feedback. How can you use exit tasks in such a way that their discussion invites collaborative thinking and being able to revise responses and paths to a task's solution?

___

___

___

___

- Based on your reading and experience with mathematics tasks, generally, how will *you* use an exit task in your classroom? Provide one or two examples of tasks that you have created or located from regularly used resources, and briefly discuss how you would use these as exit tasks.

___

___

___

___

Ponder This:

- Consider the following comment from a teacher who is also the chair of his mathematics department: *"When we first got into regularly using exit tasks, the department's biggest challenge was where to find really good tasks beyond those found in our textbooks or curriculum guide. We decided to meet for an hour or so each Monday and just pull our resources and visit websites we liked. What a great use of our time. This gave us a chance to meet informally, and now I have files of exit tasks for each of our mathematics courses, and our tasks are downloaded and adapted for classroom use every day. We still meet on Mondays, but our work is now related to 'fine-tuning' tasks we have used."* What do you think of this plan? Is this something your grade level or department would do? Share your responses.
- Challenging and engaging mathematics tasks are not always word problems! Using numbers or geometric shapes, create a task you could use as an exit task at your grade level or within a high school mathematics course you teach.

Video—Planning an Exit Task: First Grade

Video 5.1

http://bit.ly/3AYQNAw

Beth Kobett discusses the exit task and its purpose, and the importance of the time students need to become engaged in solving the task. Megan, a first-grade teacher, describes her exit task. She and her mathematics coach, Kristen, co-plan a 3-Act exit task that focuses on strategies for adding 2-digit numbers, with an emphasis on important place value concepts. Megan discusses how her students will use varied representations as they solve the task and possible responses students will demonstrate. She also mentions her use of one of the Interviews technique tools for recording the comments of her students as they share their solution strategies.

Think about one of your own classes. What did you see in this snapshot of Megan's planning for use of a 3-Act task that you might be able to plan for as you create your own 3-Act exit task for students at your grade or course level?

TOOLS FOR USING EXIT TASKS IN THE CLASSROOM

As you begin to plan for and use exit tasks in your classroom, you will think about how and when to use the exit task in your lesson and how you will regularly plan for and create or adapt a task. You will also need to validate the mathematical importance of your task, ensuring a strong connection to your lesson and its intersection with the mathematical practices or processes. Consider the following classroom challenges, classroom-based responses, and tools for planning for and using exit tasks.

1. **Think About:** Do I use the exit task every day?

Classroom Response: *In my PreK–5 school, our primary-level (PreK–2) teachers use an exit task each day. They review student performance on the tasks by taking notes as students work in small groups and then reflecting on the comments made each day. Their review of the notes they collect guides their planning. At Grades 3–5 and in our middle school, which is next door to my building, the teachers use the exit task three days a week, typically Mondays, Wednesdays, and Fridays. I am told that our high school mathematics teachers follow the same exit task schedule as our teachers in Grades 3–8. At these grade levels, the mathematics topics and the time needed for students to engage in an exit task, as well as the time the teachers need to review exit task responses, are such that using the tasks a few days a week seems to work best.*

2. **Think About:** How can I plan for using exit tasks?

Classroom Response: *I'm a third-grade teacher, and my partner teaches high school mathematics in our school district, and we have both used the Planning for Exit Tasks Tool (Figure 5.3) for several months. When we began using exit* *tasks, we found the tool to be most helpful. Now when we plan, we just refer to it, but we do provide actual comments for expected student responses to the task and always enter in our summary of the class responses and notes for planning—next steps.*

FIGURE 5.3 • **Planning for Exit Tasks Tool**

<table>
<tr><td>Date: November 9</td><td>Mathematics Standard: Write and set times using digital and analog clock times to the nearest 5 minutes using a.m. and p.m.</td><td rowspan="2">Task Level (circle one):
Lower-Level Demand (Memorization)
Lower-Level Demand (Procedures Without Connections)
Higher-Level Demand (Procedures With Connections) [circled]
Higher-Level Demand (Doing Mathematics)</td></tr>
<tr><td>Grade Level: 3</td><td>Mathematical Practices/ Processes: Problem Solving: Make sense of problems; use appropriate tools; attend to precision.</td></tr>
<tr><td colspan="3">Task: Frankie left the house at 4:15 p.m. His bus came 10 minutes later. His bus ride to school took about 25 minutes. When did the bus come? When did the bus arrive at the school? (Write the digital times and show the times using your analog clocks.)</td></tr>
<tr><td colspan="3">Expected Response:
The bus came at 4:25 p.m.; the analog clock will show the shorter hand on the 4 and the longer hand on the 5; the bus arrived at school at 4:50 p.m.; the analog clock will show the shorter hand between the 4 and 5 and the longer hand on the 10.</td></tr>
<tr><td colspan="3">Summary of Class Responses: I was pleased that most of my students were able to determine the digital time for both solutions. But several of them (between 8 and 10) were challenged by having to set the analog times. Guessing only, but I think they just use digital time more in their daily lives.</td></tr>
<tr><td colspan="3">My Planning—Next Steps: We will do activities setting analog clocks tomorrow, and I will observe and use the Interviews and Show Me techniques as needed to monitor the progress of my students.</td></tr>
</table>

A blank template version of this figure is available for download at **https://qrs.ly/wsetnnz**

3. **Think About:** What about feedback? When and how much?

Classroom Response: *One thing I learned, right off the bat, is that I need to find time to very quickly review my students' exit tasks. I always provide teacher-to-student feedback to my seventh graders. I try to both support the strengths in their work and also offer help. Time permitting, I also ask for their student-to-teacher feedback about the exit task and their progress with the task. For example, take a look at Robyn's response to an exit task that I adapted from some of our work with patterns with repeated multiplication and exponents in my Algebra 1 course (see page 164). I asked the students to use any of the following as a solution pathway for the exit task: numbers (exponents or a pattern), a table, or a graph. I also asked them to share their solution with a member of* *their group. I was able to quickly see that Robyn indicated the dollar amounts doubling in the table she prepared, but I was also able to quickly see that her table only extended to 4 days, not 5, showing $16 on day 4 and a total of $30. When I asked, "How do you know?" Robyn's shoulder partner just pointed to day 4 in the table and then the phrase "for the next 5 days" in the problem. No words! They both smiled at each other. Robyn added another row to her table, doubled 16, and added to find a total of $62, which was greater than $50. The student-to-student feedback was priceless, and my feedback to Robyn was very positive, but I couldn't resist saying, "Hey, could you have used exponents?"*

I often summarize particular feedback points (student-to-teacher, student-to-student, or teacher-to-student) and use them for interviews with a student or small group of students, which is what I will do with Robyn—just she and I.

Congratulations! You won the class prize! You can choose between the following options. Which prize would provide more money?
You get a $50 bill.
You get a $1 bill today, but each day for the next 5 days you get twice the previous day's amount of money.
Which prize do you want? Show how you solved the task. Use numbers, a table, or a graph, and write a statement explaining your solution strategy.

4. **Think About:** What about the levels of demand of my tasks?

Classroom Response: *When I first began using the exit task, I thought it was not that big of a deal. I mean, I had been using exit slips for years. But then when we began talking in grade-level groups, I came to the conclusion that my exit slips were maybe too quick and most always at a low level. Our whole school spent some time with both* Webb's (1997) *Depth of Knowledge and* Smith and Stein's (1998) *taxonomy of mathematical tasks. An activity we ended up doing, first with my fifth-grade-level team members, was to present a task to the team and then have them determine if the response demand was lower-level (memorization or procedures without connections) or higher-level (procedures with connections or doing mathematics). Later, we did this as a whole-staff activity, which was a challenge at first, because if you hadn't taught at a particular grade level, it was hard to determine the mathematical response demand of the task. Our grade-level groups now have online files, which are used regularly, that include for each task the mathematics content and mathematical practice or process standard addressed and whether it provides for lower- or higher-level demand. And, while our school district does not use the DOK (*Webb, 1997*), I have friends who use the DOK levels similarly to the way we use the* Smith and Stein (1998) *taxonomy of tasks.*

As you begin to regularly use exit tasks, it will be important to keep them organized for future use. The Exit Task Organizer Tool (Figure 5.4) can be adapted for your grade-level, department, or school team and form the basis for an online accessible file system that you can use as exit tasks are located, created, adapted, and used. Ask your teacher colleagues to complete this brief form and attach it to tasks they have found, created, or used. As they use and adapt the tasks, they can add additional notes, feedback, and even advice for administering the task.

FIGURE 5.4 • **Exit Task Organizer Tool**

Grade Level: 4	**Dates Used:** March 14	**Today's Date:** March 26
Mathematics Standard: Partition shapes into parts with equal areas. Express the area of each part as a unit fraction of the whole.		
Mathematical Practices/Processes Engaged (check those that apply): ☑ 1. Make sense of problems and persevere in solving them. ☐ 2. Reason abstractly and quantitatively. ☐ 3. Construct viable arguments and critique the reasoning of others. ☐ 4. Model with mathematics. ☑ 5. Use appropriate <u>tools</u> strategically. ☐ 6. Attend to precision. ☐ 7. Look for and make use of structure. ☐ 8. Look for and express regularity in repeated reasoning.		
Task Level (circle one): 1. Lower-Level Demand (Memorization) 2. Lower-Level Demand (Procedures Without Connections) 3. Higher-Level Demand (Procedures With Connections) (4. Higher-Level Demand (Doing Mathematics)) [circled]		
Exit Task: Shade $\frac{2}{3}$ of the area of the following rectangles in 2 different ways. Tell how you know that what you shaded is $\frac{2}{3}$.		
Exit Task's Solution: Drawings need to show 4 of 6 parts of the rectangular region shaded.		
Differentiation Decisions: Will convene a small group of those students who were unable to determine $\frac{2}{3}$ of the rectangles for an interview. Others will begin instruction involving equivalent fractions using number line representations.		
Suggestions for Exit Task Revision: Could use other shapes and fractions. For Grade 5, could consider equivalent fractions to $\frac{2}{3}$ using the number line.		
Comments: I used this with both rectangular and circular regions. (Eric; March 15)		

A blank template version of this figure is available for download at **https://qrs.ly/wsetnnz**

The three examples that follow, one at the second-grade level, one at the sixth-grade level, and one at the high school level, provide exit tasks and student responses and address everyday considerations for use of the exit tasks.

Grade 2—Understanding Place Value: Jenna checked her files and located some tasks to review for her lessons during the week on the following place value standard:

> Understand that the three digits of a three-digit number represent amounts of hundreds, tens, and ones (e.g., 857 equals 8 hundreds, 5 tens, and 7 ones).

Jenna wanted to select just the right exit task for her lesson. First, let's consider Jenna's task choices:

A.	**B.**
Julio and Mark are arguing about who has the largest number. Help them out! Show your thinking and prove who has more. **Julio** / **Mark** 3 hundreds / 5 tens 2 tens / 1 hundred 9 ones / 6 ones	Choose from the values below to create a number with a value between 300 and 600. How many combinations can you find? Use base ten blocks or draw pictures to show your work. 83 ones · 56 tens · 5 hundreds · 9 ones 2 ones and 8 tens · 27 ones 6 tens and 12 ones 32 ones and 2 hundreds 4 ones and 3 hundreds
C.	**D.**
How many groups of one hundred are in the following number? How many groups of ten? Draw a picture or use base ten blocks to show how you know. 534	Name the value of each of the digits in the following number. 797

Jenna recently wrote Task A after seeing something similar in her mathematics textbook's supplemental materials. She used both Task B and Task C last year during her place value unit and hasn't used Task D since she learned about the difference between an exit ticket and an exit task. Jenna reflected for a moment about how she used to regularly use exit tickets mostly because she thought that was how she was supposed to end every mathematics class. She knew that she gathered very little information from her students in those days and now loves how using exit tasks has provided her with a much better and deeper understanding of what her students know. As she locates or creates exit tasks, Jenna finds it helpful to consider the cognitive demand of each of her tasks. She usually likes to do this in collaboration with her teammates, so she asked Hannah and Ted to take a look. After some robust discussion, they decided that Tasks A, B, and C all represented higher levels of cognitive demand (procedures with connections).

Jenna wanted to see if her students could reason about the place value beyond just looking at the place of the digits; she had plenty of evidence that they could do this. With this in mind, she narrowed her choices down to Task A and Task B. While she really loved Task A, she decided it would be better as a hinge question.

She also felt that students should be able to answer Task A quickly, and she could assess if students were using their understanding about place value or just using order to record the value. Jenna chose Task B because she felt that the task responses could help her in assessing her students' reasoning about place value. She also liked that Task B offered great differentiation opportunities. She was curious to see how students would put together multiple combinations. Next, she knew she would need to dedicate some real time for students to mess around with this task. She would walk around while they were completing the task and take notes while the students were working. She would note when, if, and how students used manipulatives and if particular students needed additional assistance to complete the task. She was particularly interested in a few of her students who were still struggling with the use of base ten blocks. She had seen a couple of them count the ones on a ten stick several times and this concerned her, while others were able to mentally combine place values. She was excited that this task provided so many entry points. Jenna mentally checked her exit task list:

- ✓ Involves appropriate mathematics
- ✓ Is a high cognitive task
- ✓ Differentiates for multiple learners
- ✓ Promotes reasoning about place value

She was ready to go! She decided to give the exit task after a quick Show Me activity using digit cards. The pre–exit task Show Me was helpful as a verbal rehearsal for the exit task, and she has found that many students need it. She wanted to make sure students had plenty of time to show their thinking and understanding about place. She would use this information to plan some targeted lessons for individuals and groups. She wanted to know the following:

- Could her students combine multiple place values to create new numbers?
- How do her students represent their thinking?
- Can her students still recognize the value of the number when it is presented in different sequences (e.g., 2 ones and 8 tens)?
- How do her students build the values?

After providing the exit task, Jenna reflected on the students' work. Overall, she noted that most students were able to build at least two combinations. Four students demonstrated tentative understanding as they reasoned through 6 tens and 12 ones to recognize that the value was 72. While most students were able to combine values using number bonds (e.g., 60 + 12 = 72), quite a few used base ten blocks to construct the values. Three other students flew through the exit task, and she will need to consider next steps for them. Her plan was to move on to work with three-digit addition and subtraction, but after reviewing the exit task results, she decided one or two more days on place value was needed. After all, place value is a critical concept for second graders.

She took a deep breath and thought about how far she had come from a few years ago when she gave exit tickets almost every day. She knew so much more about her students' mathematical conceptual understanding and thinking now. She knew that use of the exit task was making a huge difference in her planning and teaching.

Grade 6 Statistics—Measures of Center and Range: Deon, a Grade 6 mathematics teacher, and his students had been working on understanding and applying statistics—in particular, the measures of center and variability. He monitored his students' reasoning using Observations and Show Me opportunities involving representations, which involved color tiles and graph paper drawings, as well as using estimation and equations. Deon had hoped to include the use of an exit task in today's lesson. He had recently begun to use the Open Middle website (https://www.openmiddle.com) and was eager to try the Grade 6 "Mean, Median, and Range 2" (DOK 3: Strategic Thinking) problem, submitted to the site by Eric Berchtold, Melissa Minnix, and Robert Kaplinsky (Figure 5.5).

FIGURE 5.5 • **Measures of Center and Range**

Create a set of five positive integers from 1 to 20 so that the values of their mean, median, and range are the same and have the greatest possible value. Five integers: __ __ __ __ __ Greatest possible value is: __

Source: Mean, Median, and Range 2 (https://www.openmiddle.com/mean-median-and-range-2/), from OpenMiddle.com, Eric Berchtold, Melissa Minnix, and Robert Kaplinsky © 2021. Reprinted with permission.

Deon reviewed the levels of demand described by the Smith and Stein (1998) Mathematics Task Analysis Guide (Figure 5.1). He noticed the potential for higher-level cognitive demand of the task, which:

- Required nonalgorithmic thinking.
- Required students to explore and understand the concepts related to mean, median, and range.
- Demanded self-monitoring or self-regulation of one's own cognitive processes.
- Required students to examine task constraints that may limit possible solutions and strategies (i.e., using five integers from 1 to 20, ensuring that the values of the mean, median, and range were the same).

With 12 minutes or so left in the class period, Deon introduced the exit task, giving groups of three students access to color tiles, graph paper, calculators, and a large index card to record solutions and show the group's work. Deon had reviewed the problem on the Open Middle website (www.openmiddle.com) while planning his lesson, and he observed that most of the groups were using color tiles or graph paper lengths along with estimation and computation to help determine the mean, median, and range. Deon sensed that while the class was challenged, they were also applying their understandings of mean, median, and range as they solved the exit task. As Deon collected the group solutions to the task, two of the groups indicated that they had not finished. He allowed the groups to finish the task at the end of the day and submit it to him via email or first thing in the morning. Deon reflected on the use of this particular exit task, thinking about how he might make it better or just use it differently. One consideration was to perhaps add a Show Me request involving the explicit student use of number line drawings to literally show the mean and range of the data (rather than graph paper or color tile modeling). Another thought was to assign the exit task to individual students rather than a group. Overall, Deon was very pleased with the extent to which his students were truly engaged in the task, recalling that earlier this year, several struggled to persist on tasks that required considerable cognitive effort. He thought assigning this particular task as a small-group activity may have helped keep some of his students focused and fully engaged.

High School Algebra 2—Trigonometric Functions: The students in Nasira's Algebra 2 class were working on comparing graphs of functions that rise and fall, with some containing instances that repeat, take place over certain intervals, and can be modeled with periodic functions. For example, students examined graphs of periodic functions such as the cycles of the sun and the moon and how the height of a point on a Ferris wheel moves between the same high and low.

After identifying the dependent and independent variables and discussing how the dependent variable changes in each situation, Nasira felt it was a good time to distribute the following exit task (adapted from Illustrative Mathematics, Algebra 2) on half-sheets of paper to each student.

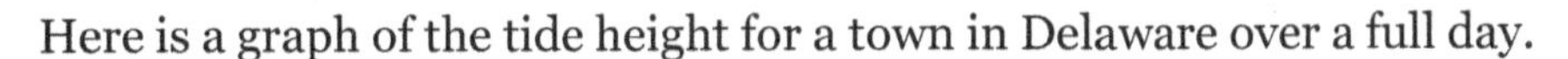
Here is a graph of the tide height for a town in Delaware over a full day.

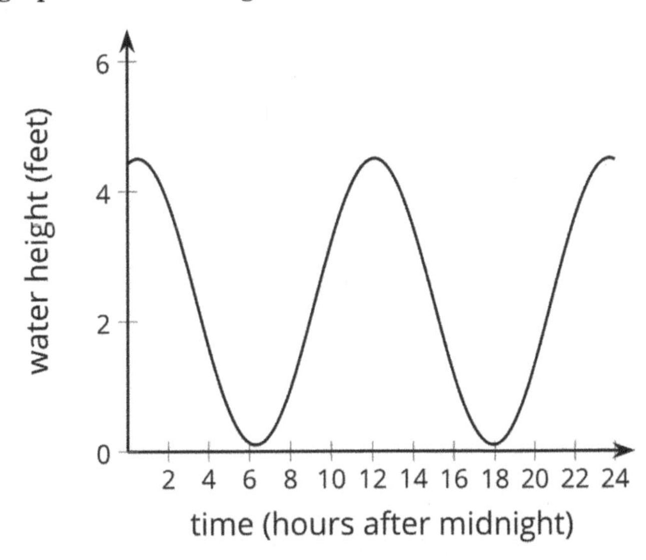

For the graphed situation, describe the dependent and independent variables. How does the dependent variable change? Be as specific as possible.

Students were given a few minutes to read over the exit task and then respond individually. Nasira prompted the students to proofread their responses—self-assessment and rehearsal time—before forming groups of three where they described their reasoning to each other. In particular, Nasira listened for responses that included descriptions related to where the function was increasing/decreasing and phrases that included *maximum* and *minimum*, *positive* and *negative*, *interval*, and *period(icty)*. During her planning period, which followed this lesson, she reviewed student responses, sorting them into those that correctly or incorrectly identified the (in)dependent variable(s) and accurately described how the dependent variable changed.

Focusing on Feedback

There are many opportunities for you and your students to engage in feedback during and after an exit task. As always, the feedback should be directly connected to the purpose of the assessment. When selecting and implementing the exit task, think about what information you need about the students' understanding as well as how you might use feedback to enhance and advance student understanding. As you plan for the implementation of the exit task, consider the many potential feedback opportunities for you and the students, the timing of those feedback opportunities, and how the feedback will be used to support your next instructional steps. First, determine the purpose of the feedback—what you want to learn about your students' mathematical understanding. Next, consider how and when you will facilitate the feedback during the exit task. Take a moment to ponder the following feedback pathways as you design and implement your next exit task.

Teacher-to-Student Feedback:

Purpose: You want to continue to support the students by providing additional feedback to deepen understanding while students are completing the task.

Implementation: The teacher monitors the students as they work on the task and may provide gentle feedback, receive questions, offer alternate solution pathways, clarify understanding, or record observations as students work. During or at the conclusion of the task, teachers might summarize solution pathways or address common questions. Teachers might also wait until the task is completed to provide individual written feedback about the entire task.

Student-to-Student Feedback:

Purpose: You want students to engage in discourse and believe that they could strengthen their understanding by sharing strategies and solution pathways.

Implementation: Among all of the Formative 5 techniques, Exit Tasks might provide the most opportunities for students to provide robust feedback to one another. For example, students could complete a rough draft of the task on their own and then work with a partner to compare solution pathways, strategies, and mathematical conclusions. After student-to-student feedback is shared, students revise their tasks and share how the feedback influenced their revision and supported new mathematical understanding.

Student-to-Teacher Feedback:

Purpose: You want to learn about the students' perception of the task you selected and the mathematical expectations of the task.

Implementation: The teacher asks students for feedback about the exit task. The students voice their opinions about the task including what they liked and didn't like and even, perhaps, make suggestions about changes that could be made to the task. Teachers might ask:

- Were you able to show your understanding of _______?
- Why or why not? What did you like about this task? Why?
- How do you think the task could be improved?

Now, it's your turn:

1. Select an exit task that you are going to implement. Include the task here:

2. Decide on the purpose of the feedback that you want to collect from this exit task.
3. Next, record the questions and prompts for the task you selected, for each type of feedback.

Teacher-to-Student Feedback:

Student-to-Student Feedback:

Student-to-Teacher Feedback:

After implementing the task, reflect on the feedback.

(Continued)

(Continued)

How will the feedback change your instruction?

What surprised you?

What are your next steps?

TECHNOLOGY TIPS AND TOOLS FOR EXIT TASKS

There are several websites mentioned throughout this module with mathematical tasks that hold good potential for use as exit tasks. There are also some promising digital tools that provide teachers with the opportunity to pose, capture, and analyze student responses to exit tasks. Many of those mentioned in the Show Me (Module 3) and Hinge Questions (Module 4) modules have potential use with Exit Tasks too (e.g., CueThink, Google Forms, Formative, Padlet). Another popular tool is Socrative (www.socrative.com), which, among other features, allows teachers to choose or develop exit tasks, view live student responses, receive questions from students, discuss results, or ask more follow-up questions.

The power in using exit tasks with or without digital technology is in how the information collected about student performance is used. A true benefit of using digital tools with the Formative 5 techniques is having the ability to capture, share, compare, and review responses. The convenience of being able to seamlessly analyze and rewind, using the technology as a means to efficiently provide students with meaningful feedback, is so important to your planning and teaching. Opportunities afforded through using technology often provide your class with the ability to see all student responses, giving your students a chance to provide each other with feedback, learn from a mistake, and share understandings with each other.

USING EXIT TASKS IN *YOUR* CLASSROOM

As with each of the Formative 5 techniques, considerations for use of the exit task will be an important element of your lesson planning. The exit task caps major concepts or skills presented within your lesson, and as noted previously, the exit task and hinge question provide mathematical closure to your lesson. As with observations, interviews, Show Me, and the hinge question, student responses to the exit task help to identify their instructional needs. As you plan for the next day's lesson, include considerations related to grouping and differentiation. Related to the more individualized or small-group Show Me technique, the exit task provides a sampling of actual student performance. The more experience you have with the use of the exit task, the less time it will take you to create, adapt, or locate viable exit tasks; present and review them; and provide feedback to your students. However, regular use of the exit task includes the following time-demanding elements:

- Locating or creating a task related to your mathematics lesson; considering the mathematics content, the mathematical practices or processes engaged, the level of demand of the task, and the time needed for your students to complete the task
- Presenting the exit task to your students
- Reviewing and analyzing student responses to the exit task
- Providing feedback to your students based on their response to the exit task as well as providing an opportunity for students to provide feedback to you related to the exit task experience
- Adapting your planning and teaching based on exit task responses

Steve: I use the exit task most days. The sixth- and seventh-grade teaching teams at our intermediate school often plan our exit tasks together. While we plan, we spend a lot of time discussing particular challenges math-wise of the lessons we are preparing. We all agree that both our hinge questions and exit tasks need to focus on the really important mathematical ideas within our lessons. Whenever I do my final "dry run" of the exit task for my lesson, I think about the extent to which my students should be ready for the exit task as well as consider possible student responses. Both of our teaching teams hope that the exit tasks we provide engage our students in, for the most part, higher-level thinking that is all about reasoning and problem solving. More recently, we have stored our exit tasks in an online folder—organized by mathematics content topic. As we access and then use one of our team's exit tasks, we make comments about their task, suggesting changes, if appropriate, using the Exit Task Organizer Tool (see Figure 5.4). Regularly adding to and updating our exit task online folder has helped us in building a collection of tasks that we can use, along with other resources and online tasks that we have added and can repurpose for our classrooms. To some extent, our mathematics teaching has always included problems or tasks. But now we know what to consider when selecting or creating exit tasks and how to actually use the exit task to advise our planning and teaching.

The tools presented in the previous section of this module and the teacher comments and examples provided should be helpful to you as you consider your own planning for and use of the exit task. The exit task is your end-of-lesson document of progress and provides mathematical closure for a lesson or perhaps a multiday topic. It provides a record for future use, whether updating or comparing progress from time to time, and a work sample you can use to consider grade-level differentiation efforts or share at parent conferences. Perhaps more importantly, it adds to all you know, through your everyday use of the Observations, Interviews, Show Me, and Hinge Questions techniques, about your students' mathematical knowledge and background.

Time to Try

Like the hinge question, your use of the exit task will often serve as a closure activity for the lesson. And, like the Show Me technique, the exit task is a performance-based assessment.As you plan, you will need to find the time to create or locate an exit task identified as being a higher-level demand task—one that engages procedures with connections or involves your students in doing mathematics (Smith & Stein, 1998). One way to think about an exit task is that such a task focuses on important mathematics of the day or perhaps several days and engages your students in doing that mathematics. You may decide to have all your students complete the exit task or perhaps have your students work in small groups to solve the exit task. As you engage students in the exit task, you provide and assess their responses, and you may provide opportunities for student-to-student and student-to-teacher feedback, as well as providing your feedback to the students. Due to the time related to reviewing exit task performance, you may decide to use exit tasks several times a week, rather than every day. It's now time to try this technique! Identify a mathematics lesson that will occur tomorrow or in the next few days. Using the following table, provide brief responses to each of the statements related to planning for and actual use of exit tasks. Discuss your exit task "plan" responses with a colleague.

STATEMENTS	RESPONSES
Mathematics/Lesson Focus	
Anticipation: *Specifically, what mathematics does the proposed exit task emphasize?* *Is the exit task set within a particular context (e.g., time zones; the aquarium)? If so, identify the context.*	
Using the Planning for Exit Tasks Tool (Figure 5.3) to help in your planning, provide a draft of an exit task.	

STATEMENTS	RESPONSES
How would you use or adapt the Exit Task Organizer Tool (Figure 5.4)?	
If your students work in small groups to complete a lesson's exit task, how could you use Show Me responses to the task as a way to engage student-to-student feedback? *How will you summarize student and class responses on your exit task?*	

Video—Using Differentiated Exit Tasks: Third Grade

Video 5.3

http://bit.ly/3umb2Ul

Beth Kobett discusses how Brooke, a third-grade teacher, designed her exit task to reflect the needs of her diverse classroom. The task was to design a zoo, with students drawing exhibits for various animals, using skills and understandings related to perimeter and area. Brooke scaffolded questions as she met with students individually and in small groups, as a way to differentiate as well as engage her students. As Brooke monitored the progress of her students in solving the exit task, she used the Observations, Interviews, and Show Me techniques. Brooke felt that her task was problem-based and provided varied routes for students to demonstrate their understanding.

Think about and discuss how you monitor the performance of your students as they engage in and solve mathematics tasks. How do you differentiate to fully engage your students in such tasks? Do you scaffold questions? Do you engage students in groups? Do you provide access to varied representation tools? Discuss responses with a grade-level or mathematics department team.

SUMMING UP

The exit task serves as an end-of-lesson barometer of the mathematical understandings of your students. Depending on the focus of the lesson, the exit task may assess a full range of mathematical expectations. However, such tasks will more frequently address problem solving and reasoning. Reviewing student responses to your exit task should provide you with direction for your planning and instructional next steps. Providing opportunities for teacher-to-student, student-to-teacher, and student-to-student feedback will also be an important component of your planning for and use of the exit task. Monitoring the success of your exit tasks should provide you with a regularly updated and growing collection of validated tasks for future years. And, as with the other elements of the Formative 5, the continuing use of exit tasks is enhanced when school or grade-level learning communities work together in their creation, use, and revision. This, the final technique of the Formative 5, completes the palette of classroom-based formative assessment techniques you will use to both monitor student progress and guide your planning and teaching every day.

Your Turn

Rate, Read, Reflect! Consider and respond to the following questions with your grade-level or mathematics department teaching team or with teams across multiple grade levels.

1. How frequently will you use the Exit Tasks technique? (check one)

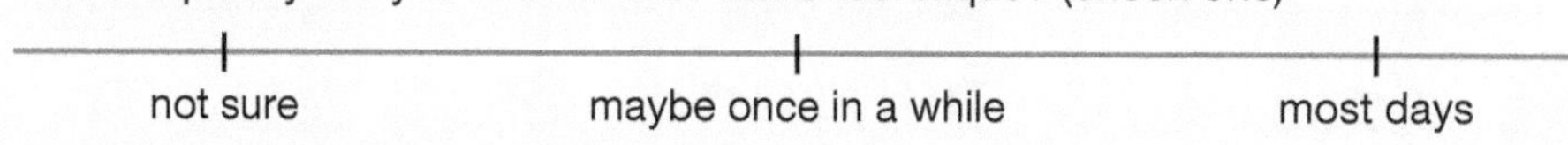

2. As you consider use of exit tasks, how will you analyze student responses to the task? (check all that apply)

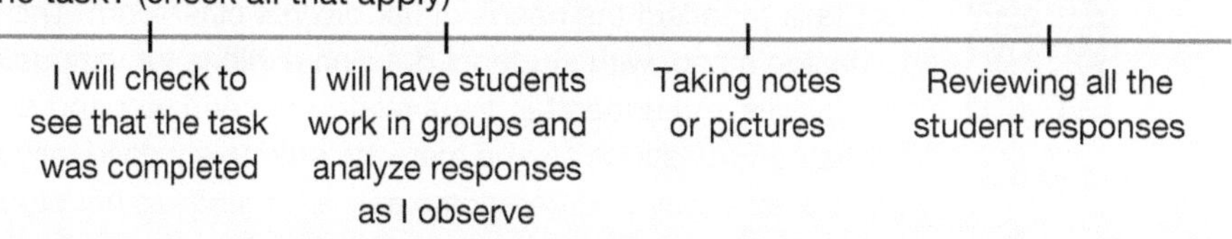

3. As you consider planning for and implementing an exit task, what do you see as your biggest challenge? (check all that apply)

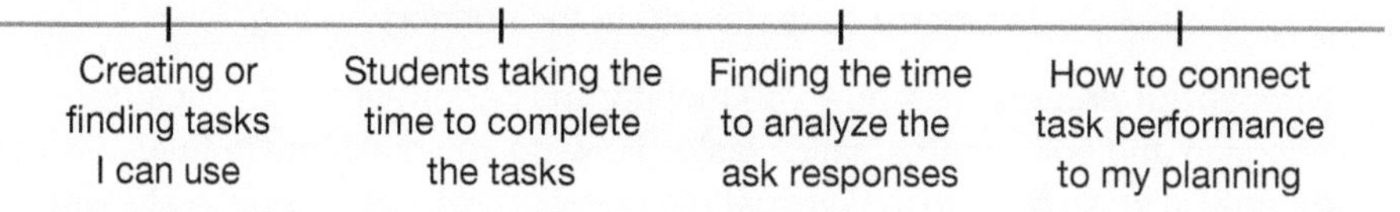

4. As you think about regular use of exit tasks, identify several sources of tasks that may help you in planning for and creating exit tasks.
5. How will you provide time for your students to engage in an exit task and for you to perhaps use the Observations and Show Me techniques to monitor student progress on the exit task?
6. How could you use exit task responses or perhaps a summary of such responses to communicate student progress in mathematics at parent–teacher conferences?
7. How could you and colleagues create a plan to organize the exit tasks used by your grade level or department online? How would you update such a filing system?
8. How will you find the time to provide teacher-to-student feedback on exit tasks you provide? In writing as you review the task solutions? Orally as students work individually or in small groups?
9. Consider two or three major mathematics topics at a particular grade or course level. Create a draft of an exit task for each of these topics. Share the draft tasks with colleagues and, if time allows, try the tasks out with your students.
10. Professional Connection: Consider reading "Three Ways to Enhance Tasks for Multilingual Learners" by Sarah H. Roberts, Zandra de Araujo, Craig Wiles, and William Zahner (2022) in *Mathematics Teacher: Learning and Teaching PK–12* 115, no. 7, available from https://pubs.nctm.org/view/journals/mtlt/115/7/article-p458.xml, then discuss the following:
    a. Roberts et al. (2022) discuss three strategies for enhancing mathematics tasks to make them more accessible for multilingual students: (1) using and connecting multiple representations; (2) thinking through language obstacles; and (3) contextualizing mathematics concepts and problem-solving activities to further students' mathematical learning alongside their language development. How would you use these strategies in your own classroom to make mathematics tasks more accessible?
    b. Consider the following statement from Roberts et al. (2022): "If a task includes only a single representation, one way to enhance that task for use with multilingual learners is to incorporate one or more additional representations" (p. 459). Think about an exit task you may have used. How could you adapt the task to include the use of multiple representations?

EPILOGUE

NOW WHAT?

If you are reading this page, we would like to believe that you have read and completed many activities from *The Formative 5 in Action*, used or adapted the book's tools for each of the Formative 5 techniques, viewed the classroom videos provided within each of the book's modules, and discussed your use of the Observations, Interviews, Show Me, Hinge Questions, and Exit Tasks techniques with others—in your grade-level team or mathematics department. But keep reading! As noted in the Preface to this book, *The Formative 5 in Action* is more than a revision; this is essentially a different book, and includes a full range of activities along with a very specific focus on the importance of feedback and its connection to the classroom-based formative assessment techniques we call the Formative 5. Importantly, *The Formative 5 in Action* is organized to fully support teachers at the K–12 level. We also hope that you have noticed and appreciated our inclusion of comments from scores of K–12 classroom teachers, mathematics coaches/specialists, and other teacher leaders as they have encountered and addressed the challenges related to using a particular Formative 5 technique in their own classroom-based settings. Deeply analyzing the impact of classroom-based formative assessment and then creating, testing, and actually determining the techniques that have become the Formative 5 has been a time-consuming, challenging, and rewarding effort. But let's not get ahead of ourselves. As you prepare to engage in implementing the Formative 5 on a regular basis, the following points, related to each of the techniques, may be helpful to you. Take a look.

Observations

- Recognize that observations you make throughout your day inform your teaching minute by minute. What's observed may cause you to shift a lesson's focus, alter the pace of the lesson, engage particular resources, and more. How you anticipate and use observations provides you with that initial link to how formative assessment can and will influence your planning and teaching.
- You observe all the time. So, as you plan a lesson, anticipate what you may observe and how you will record your observations. Also, be sure to consider how you may provide feedback to your students, engage them in providing feedback to you, or plan for students to provide feedback to each other.

Interviews

- Interview questions and student responses provide you with an everyday opportunity to deepen your understanding of what your students know. However, this is not a technique to be used primarily to assess what may appear to be deficiencies. Think of your student interview responses as an opportunity to "spot strengths." We have found that the interview also provides an opportunity to explore unique, creative, and advanced understandings and solutions to problems as well as assess student mathematical dispositions.
- Your interviews are neither long nor formal. We all know how aware students are when they sense something is different or formal. Your interview is a brief, directed conversation where your focus is on listening to understand student thinking. Many interviews are simply, "Tell me what you are doing and why you are doing it that way." And, yes, student responses to such prompts are wonderful examples of student-to-teacher feedback.

Show Me

- You will use Show Me prompts to extend what you have observed. It is very natural for you to ask your students to show what they are doing and, as appropriate, offer an explanation of the solution strategies they have used. Student responses to your Show Me prompt provide you with student-to-teacher feedback, and also present instructional decision-making opportunities as to whether you continue or change the direction of your lesson.
- Your decision to use the Show Me technique begins with the planning of a lesson. As you consider what you will observe, or aspects of the lesson's mathematical engagement that may warrant an interview, will there also be places within your lesson where you may want to just stop whatever you are doing and ask your students to show what they are doing, and perhaps discuss why? While your actual use of the Show Me technique is most likely spontaneous, its use is often a response to an observation or interview. We sometimes think of Show Me as an abbreviated performance-based interview.

Hinge Questions

- The hinge question seeks to assess the progress of your class within a lesson. It's not a discussion question. It's diagnostic in that you are asking students about a major element of the day's lesson and seeking responses that will determine what your students know or understand. Some think of the hinge question as a whole-class interview. The student responses, in a sense, assess your lesson and provide direct student feedback to you as to your next steps planning-wise and instructionally.

- When you plan a lesson, it is never finished and is always "under construction." In anticipating and even providing a hinge-point question within your lesson, it's always your responsibility to adapt the question as needed. This could include changing the language and focus of your question. So, as you teach, be prepared to revise your hinge questions as needed.

Exit Tasks

- Like Show Me, responses to an exit task provide a sampling of student performance, and often provide mathematical closure to your lesson. Plan for the time needed to review student responses and provide teacher-to-student feedback to your students or engage your students in providing student-to-student feedback to exit task responses. Given the time needed to review exit task performance and provide feedback, you may decide to use exit tasks a few days each week rather than every day. Exit task users have told us that collecting student exit tasks over time provided them with a trajectory of their students' mathematical understandings and accomplishments during the year.
- Issues related to accessing and using appropriate exit tasks are often expressed as a concern related to the regular use of exit tasks. While you may enjoy creating or adapting your own exit tasks, there are many websites and related digital tools that you could use as resources for exit tasks. We have also found that many school-based learning communities create grade-level or mathematics course resource files for sharing and regularly updating their exit tasks.

WHAT'S NEXT?

First, our hope is that you have discussed and engaged in the Formative 5 techniques to the point where you are now ready to use them every day in your classroom. Should you start by using each technique every day? In our experience supporting teachers, schools, and school districts with their implementation of the Formative 5, we have found that a gradual induction to implementing the techniques just makes sense. Full implementation of all the techniques will take time. As one principal noted: "The teachers in our building are beginning to understand the shift to using the Formative 5 and making classroom-based formative assessment more meaningful. With more time, I think the impact will be more visible." We have also found that some techniques appear to be easier to understand and implement than others. Consider the following comments:

> "It will take my teachers a while to become proficient with creating and using hinge questions."
>
> "We are still working on developing and locating better exit tasks."
>
> "When teachers use interviews and Show Me, our students are more comfortable with explaining their thinking."

The Formative 5 in Action has presented our plan for your understanding of and engagement with classroom-based formative assessment techniques that work. But please recognize and consider the importance of the following:

- **Time.** All teachers need time to think about and seriously consider the everyday connection between planning, teaching, and assessing.
- **Anticipation.** As you plan, anticipating student responses to activities within a day's mathematics lesson will help determine your use of the Formative 5 techniques.
- **Feedback.** Recognize and regularly think about how directly feedback is connected to what, when, and how you will use the Formative 5 techniques.

Throughout *The Formative 5 in Action*, we have discussed and visually referenced an artist's palette as one way to think about, visualize, and even apply each of the Formative 5 techniques. Well, now it's time for you to create assessment artistry—in your classroom. Next is now! You can do this. Go for it!

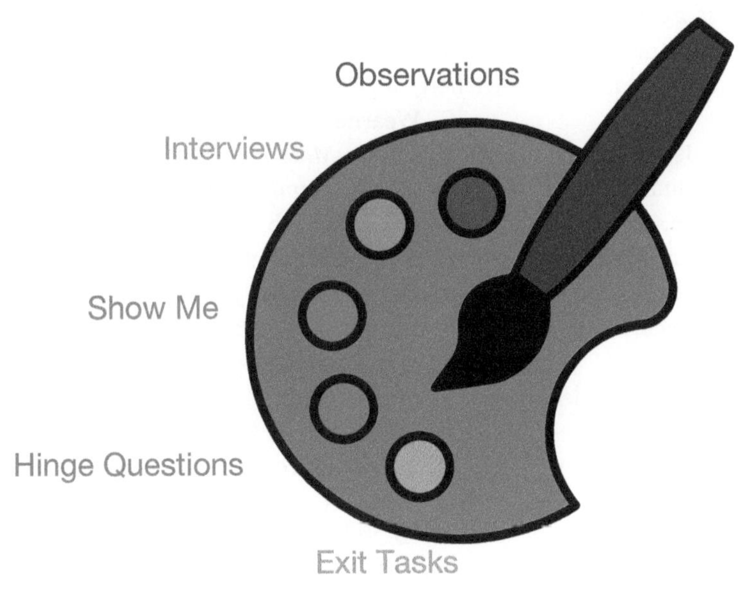

Source: iStock.com/Turac Novruzova

References

Black, P., & Wiliam, D. (1998). Assessment and classroom learning. *Assessment in Education: Principles, Policy & Practice, 5*(1), 7–74.

Black, P., & Wiliam, D. (2009). Developing the theory of formative assessment. *Educational Assessment, Evaluation and Accountability, 21*(1), 5–31.

Bloom, B. S. (1969). Some theoretical issues relating to educational evaluation. In H. G. Richey & R. W. Tyler (Eds.), *Educational evaluation: New roles, new means, part 2* (Vol.68, pp. 26–50). University of Chicago Press.

Chappuis, S., Commodore, C., & Stiggins, R. (2018). *Balanced assessment systems: Leadership, quality, and the role of classroom assessment*. Corwin.

Council of Chief State School Officers. (2018). *Revising the definition of formative assessment*. Council of Chief State School Officers.

Doyle, W. (1988, February). Work in mathematics classes: The context of students' thinking during instruction. *Educational Psychologist, 23*, 167–180.

Darling-Hammond, L. (1994). Performance-based assessment and educational equity. *Harvard Educational Review, 64*(1), 5–31.

Driscoll, M., Nikula, J., & DePiper, J. N. (2016). *Mathematical thinking and communication: Access for English learners*. Heinemann.

Fennell, F. (1998). A through the lens look at moments in classroom assessment. In G. Bright & J. M. Joyner (Eds.), *Classroom assessment in mathematics* (pp. 161–166). United Press of America.

Fennell, F. (2011). All means all. In F. Fennell (Ed.), *Achieving fluency: Special education and mathematics* (pp. 1–14). National Council of Teachers of Mathematics.

Fennell, F. (2020). Then: Assessment—So much more than tests, testing, and accountability. *Mathematics Teacher: Learning and Teaching PK–12, 113*(8), 673–675.

Fennell, F., Kobett, B., & Wray, J. (2015). Classroom-based formative assessments: Guiding teaching and learning. In C. Suurtamm (Ed.) & A. McDuffie (Series Ed.), *Annual perspectives in mathematics education: Assessment to enhance teaching and learning* (pp. 51–62). National Council of Teachers of Mathematics.

Fennell, F., Kobett, B. M., & Wray, J. A. (2017). *The Formative 5: Everyday assessment techniques for every math classroom*. Corwin.

Fisher, D., & Frey, N. (2004). *Improving adolescent literacy: Strategies at work*. Pearson Prentice Hall.

Freudenthal, H. (1973). *Mathematics as an educational task*. Springer.

Fujii, T. (2020). Misconceptions and alternative conceptions in mathematics education. In S. Lerman (Ed.), *Encyclopedia of mathematics education* (2nd ed., pp. 625–627). Springer.

Ginsburg, H. P. (1997). *Entering the child's mind: The clinical interview in psychological research and practice*. Cambridge University Press.

Ginsburg, H. P., & Dolan, A. O. (2011). Assessment. In F. Fennell (Ed.), *Achieving fluency: Special education and mathematics* (pp. 85–103). National Council of Teachers of Mathematics.

Hattie, J., & Timperley, H. (2007). The power of feedback. *Review of Educational Research, 77*(1), 81–112.

Hiebert, J., Carpenter, T. P., Fennema, E., Fuson, K. C., Wearne, D., Hanlie, H., Olivier, A., & Human, P. (1997). *Making sense: Teaching and learning mathematics with understanding*. Heinemann.

Jacobs, V. R., Lamb, L. L., & Philipp, R. A. (2010). Professional noticing of children's mathematical thinking. *Journal for Research in Mathematics Education, 41*(2), 169–202.

Jansen, A. (2020). *Rough draft math: Revising to learn*. Stenhouse.

Karp, K. S., Bush, S. B., & Dougherty, B. J. (2014). 13 rules that expire. *Teaching Children Mathematics, 21*(1), 18–25.

Kobett, B., & Karp, K. (2020). *Strengths-based teaching and learning in mathematics*. Corwin.

Kobett, B., Fennell, F., Karp, K., Andrews, D., & Mulroe, S. (2021). *Classroom-ready rich math tasks*. Corwin.

Leahy, S., Lyon, C., Thompson, M., & Wiliam, D. (2005). Classroom assessment: Minute-by-minute and day-by-day. *Educational Leadership, 63*(3), 18–24.

Leighton, J. P. (2019). Students' interpretation of formative assessment feedback: Three claims for why we know so little about something so important. *Journal of Educational Measurement, 56*, 793–814.

National Council of Teachers of Mathematics. (1991). *Professional standards for teaching mathematics*. National Council of Teachers of Mathematics.

National Council of Teachers of Mathematics. (1995). *Assessment standards for school mathematics*. National Council of Teachers of Mathematics.

National Council of Teachers of Mathematics. (2000). *Principles and standards for school mathematics*. National Council of Teachers of Mathematics.

National Council of Teachers of Mathematics. (2014). *Principles to actions: Ensuring mathematics success for all*. National Council of Teachers of Mathematics.

National Council of Teachers of Mathematics. (2018). *Catalyzing change in high school mathematics: Initiating critical conversations*. National Council of Teachers of Mathematics.

National Governors Association Center for Best Practices & Council of Chief State School Officers. (2010). *Common Core State Standards for mathematics*. https://learning.ccsso.org/wp-content/uploads/2022/11/Math_Standards1.pdf

National Mathematics Advisory Panel. (2008). *Foundations for success: The final report of the National Mathematics Advisory Panel*. U.S. Department of Education.

Nelson, H. (2013). *Testing more, teaching less: What America's obsession with student testing costs in money and lost instructional time*. American Federation of Teachers.

Polya, G. (1945). *How to solve it*. Princeton University Press.

Resnick, L. (1987). *Education and learning to think*. National Academy Press.

Roberts, S. A., de Araujo, Z., Willey, C., & Zahner, W. (2022). Three ways to enhance tasks for multilingual learners. *Mathematics Teacher: Learning and Teaching PK–12, 115*(7), 458–467. https://pubs.nctm.org/view/journals/mtlt/115/7/article-p458.xml

Scriven, M. (1967). The methodology of evaluation. In R. W. Tyler, R. M. Gagné, & M. Scriven (Eds.), *Perspectives of curriculum evaluation* (Vol.1, pp. 39–83). RAND.

Shavelson, R. J., Baxter, G. P., & Pine, J. (1992). Performance assessments: Political rhetoric and measurement reality. *Educational Researcher, 21*(4), 22–27.

Smith, M. S., & Stein, M. K. (1998). Selecting and creating mathematical tasks: From research to practice. *Mathematics Teaching in the Middle School, 3*(5), 344–349.

Smith, M. S., & Stein, M. K. (2018). *5 practices for orchestrating productive mathematics discussions*. National Council of Teachers of Mathematics.

Spangler, D. A., Kim, J., Cross, D., Kilic, H., Iscimen, F., & Swanagan, D. (2014). Using rich tasks to promote discourse. In K. Karp (Ed.) & A. R. McDuffie (Series Ed.), *Annual perspectives in mathematics education 2014: Using research to improve instruction* (pp. 97–104). National Council of Teachers of Mathematics.

Stacey, K. (2005). Travelling the road to expertise: A longitudinal study of learning. In H. L. Chick & J. L. Vincent (Eds.), *Proceedings of the 29th conference of the International Group for the Psychology of Mathematics Education* (Vol.1, pp. 19–36). PME.

Stein, M. K., Grover, B., & Henningsen, M. (1996, Summer). Building student capacity for mathematical thinking and reasoning: An analysis of mathematical tasks used in reform classrooms. *American Educational Research Journal, 33*, 455–488.

Stein, M. K., Lane, S., & Silver, E. (1996, April). *Classrooms in which students successfully acquire mathematical proficiency: What are the critical features of teachers' instructional practice?* [Paper presentation]. American Educational Research Association Annual Meeting, New York.

Stein, M. K., Smith, M. S., Henningsen, M., & Silver, E. A. (2009). *Implementing standards-based mathematics instruction: A casebook for professional development* (2nd ed.). Teachers College Press.

Stiggins, R. J. (2005). From formative assessment to assessment FOR learning: A path to success in standards-based schools. *Phi Delta Kappan, 87*(4), 324–328.

Sueltz, B. A., Boynton, H., & Sauble, I. (1946). The measurement of understandings in elementary school mathematics. In W. Brownell (Ed.), *Measurement of understanding: 45th yearbook of the National Society for the Study of Education, part 1* (pp. 138–156). University of Chicago Press.

Timperley, H. S., & Wiseman, J. (2002). *The sustainability of professional development in literacy*. New Zealand Ministry of Education.

Understanding formative assessment: A special report. (2015, November 9). *Education Week*. http://www.edweek.org/ew/collections/understanding-formative-assessment-special-report/

Watson, A., & Barton, B. (2011). Teaching mathematics as the contextual application of mathematical modes of enquiry. In T. Rowland & K. Ruthven (Eds.), *Mathematical knowledge in teaching* (Mathematics Education Library, Vol. 50, pp. 65–82). Springer, Dordrecht. https://doi.org/10.1007/978-90-481-9766-8_5

Weaver, F. J. (1955). Big dividends from little interviews. *Arithmetic Teacher, 2*(2), 40–47.

Webb, N. (1997). *Research monograph number 6: Criteria for alignment of expectations and assessments on mathematics and science education.* Council of Chief State School Officers.

Wiliam, D. (2011). *Embedded formative assessment.* Solution Tree Press.

Wiliam, D. (2018). *Embedded formative assessment* (2nd ed.). Solution Tree Press.

Wiliam, D., & Leahy, S. (2015). *Embedding formative assessment: Practical techniques for K–12 classrooms.* Learning Sciences International.

Wiliam, D., & Thompson, M. (2008). Integrating assessment with instruction: What will it take to make it work? In C. A. Dwyer (Ed.), *The future of assessment: Shaping teaching and learning* (pp. 53–84). Routledge.

Wilkerson, T. (2022, June). *Using formative assessment effectively* [President's message]. National Council of Teachers of Mathematics. https://www.nctm.org/News-and-Calendar/Messages-from-the-President/Archive/Trena-Wilkerson/Using-Formative-Assessment-Effectively/

Willingham, D. T. (2009). *Why don't students like school? A cognitive scientist answers questions about how the mind works and what it means for your classroom.* Jossey-Bass.

Yang, Z., Yang, X., Wang, K., Zhang, Y., Pei, G., & Xu, B. (2021). The emergence of mathematical understanding: Connecting to the closest superordinate and convertible concepts. *Frontiers in Psychology, 12.* https://doi.org/10.3389/fpsyg.2021.525493

Index

JENNIFER M. BAY-WILLIAMS, JOHN J. SANGIOVANNI, ROSALBA SERRANO, SHERRI MARTINIE, JENNIFER SUH, C. DAVID WALTERS

Because fluency is so much more than basic facts and algorithms.

Grades K–8

ROBERT Q. BERRY III, BASIL M. CONWAY IV, BRIAN R. LAWLER, JOHN W. STALEY, COURTNEY KOESTLER, JENNIFER WARD, MARIA DEL ROSARIO ZAVALA, TONYA GAU BARTELL, CATHERY YEH, MATHEW FELTON-KOESTLER, LATEEFAH ID-DEEN, MARY CANDACE RAYGOZA, AMANDA RUIZ, EVA THANHEISER

Learn to plan instruction that engages students in mathematics explorations through age-appropriate and culturally relevant social justice topics.

Early Elementary, Upper Elementary, Middle School, High School

JOHN J. SANGIOVANNI, SUSIE KATT, LATRENDA D. KNIGHTEN, GEORGINA RIVERA, FREDERICK L. DILLON, AYANNA D. PERRY, ANDREA CHENG, JENNIFER OUTZS

Actionable answers to your most pressing questions about teaching elementary and secondary math.

Elementary, Secondary

SARA DELANO MOORE, KIMBERLY RIMBEY

A journey toward making manipulatives meaningful.

Grades K–3, 4–8

CORWIN
A SAGE Publishing Company